BASIC CONCEPTS IN CHEMISTRY

Heterocyclic Chemistry

M. SAINSBURY

University of Bath

**WILEY-
INTERSCIENCE**

RS•C

ROYAL SOCIETY OF CHEMISTRY

Library of Congress Cataloging-in-Publication Data:
Library of Congress Cataloging-in-Publication Data is available.
ISBN: 0-471-28164-6

Typeset in Great Britain by Wyvern 21, Bristol
Printed and bound by Polestar Wheatons Ltd, Exeter

10 9 8 7 6 5 4 3 2 1

Preface

This book provides a concise, yet thorough, introduction to the vast subject of heterocyclic chemistry by dealing only with those compounds containing a single heteroatom. By restricting the discussion to these, the most important classes of heterocycles, a balanced treatment is possible, allowing the student to rapidly understand the importance of heterocyclic compounds in general to mankind and at the same time stimulating an interest in the challenges this chemistry presents.

The contents of the book are carefully designed to meet the needs of undergraduate students in the 2nd year of a degree course in Chemistry or Biochemistry and are based upon the author's own lectures given to students at Bath. Although primarily an undergraduate text, the main principles that govern heterocyclic chemistry as a whole are addressed in this book, providing a sure foundation for those wishing to widen their interest in heterocyclic chemistry in later years.

Malcolm Sainsbury
Bath

BASIC CONCEPTS IN CHEMISTRY

EDITOR-IN-CHIEF

Professor E W Abel

EXECUTIVE EDITORS

Professor A G Davies
Professor D Phillips
Professor J D Woollins

EDUCATIONAL CONSULTANT

Mr M Berry

This series of books consists of short, single-topic or modular texts, concentrating on the fundamental areas of chemistry taught in undergraduate science courses. Each book provides a concise account of the basic principles underlying a given subject, embodying an independent-learning philosophy and including worked examples. The one topic, one book approach ensures that the series is adaptable to chemistry courses across a variety of institutions.

TITLES IN THE SERIES

Stereochemistry *D G Morris*
Reactions and Characterization of Solids
  S E Dann
Main Group Chemistry *W Henderson*
d- and f-Block Chemistry *C J Jones*
Structure and Bonding *J Barrett*
Functional Group Chemistry *J R Hanson*
Organotransition Metal Chemistry *A F Hill*
Heterocyclic Chemistry *M Sainsbury*
Thermodynamics and Statistical Mechanics
  J M Seddon and J D Gale
Basic Atomic and Molecular Spectroscopy
  J M Hollas

Further information about this series is available at www.wiley.com/go/wiley-rsc

Contents

1 Introduction to Heterocyclic Chemistry   **1**

1.1   Coverage   1
1.2   Nomenclature   1
1.3   Importance to Life and Industry   4
1.4   General Principles   6

2 Pyridine   **18**

2.1   Resonance Description   18
2.2   Electrophilic Substitution   19
2.3   Pyridine *N*-Oxides   22
2.4   Nucleophilic Substitution   23
2.5   Lithiation   28
2.6   Methods of Synthesis   28
2.7   Commonly Encountered Pyridine Derivatives   29
2.8   Reduced Pyridines   36

3 Benzopyridines   **42**

3.1   Introduction   42
3.2   Quinoline   43
3.3   Isoquinoline   50

4 Pyrylium Salts, Pyrans and Pyrones   **58**

4.1   Introduction   58
4.2   Pyrylium Salts   59

4.3   Pyran-2-ones (α-Pyrones)   61
4.4   Pyran-4-ones (γ-Pyrones)   63
4.5   Reduced Pyrans   65
4.6   Saccharides and Carbohydrates   65

**5   Benzopyrylium Salts, Coumarins,
      Chromones, Flavonoids and Related
      Compounds   68**

5.1   Structural Types and Nomenclature   68
5.2   Coumarins   70
5.3   Chromones (Benzopyran-4-ones)   72

**6   Five-membered Heterocycles containing
      One Heteroatom: Pyrrole, Furan and
      Thiophene   77**

6.1   Pyrrole   77
6.2   Furan   85
6.3   Thiophene   91

**7   Benzo[*b*]pyrrole, Benzo[*b*]furan and
      Benzo[*b*]thiophene   97**

7.1   Indole (Benzo[*b*]pyrrole)   97
7.2   Benzo[*b*]furan and Benzo[*b*]thiophene   110

**8   Four-membered Heterocycles containing a
      Single Nitrogen, Oxygen or Sulfur Atom   115**

8.1   Azete, Azetine and Azetidine   115
8.2   Oxetene and Oxetane   121
8.3   Thietene and Thietane   122

Answers to Problems   125

Subject Index   141

1

Introduction to Heterocyclic Chemistry

Aims

By the end of this chapter you should understand:

- Why heterocyclic chemistry is so important to mankind
- The names of a few commonly encountered heterocyclic compounds
- Some of the major factors that govern the shape and stability of heterocyclic compounds

1.1 Coverage

The subject of heterocyclic chemistry is vast, so in this book the focus is on the more common four-, five- and six-membered systems containing one heteroatom. Little attempt is made to extend the coverage to more complex heterocycles, so that students interested in extending their knowledge will need to consult more advanced works. Fortunately, there is a very wide choice; excellent texts include *Heterocyclic Chemistry* by Gilchrist[1] and *Heterocyclic Chemistry* by Joule and Mills.[2] In addition, there are many authoritative compilations that deal with heterocyclic chemistry in much more depth.[3-6]

1.2 Nomenclature

Students will be familiar with carbocyclic compounds, such as cyclo-hexane and benzene, that are built from rings of carbon atoms. If one or more of the carbon atoms is replaced by another element, the product is a heterocycle. Multiple replacements are commonplace, and the elements involved need not be the same. The most common are oxygen, sulfur or nitrogen, but many other elements can function in this way, including boron, silicon and phosphorus. Chemists have been working with heterocycles for more than two centuries, and trivial names were often applied long before the structures of the compounds were known. As a result, many heterocycles retain these names; a selection of common five- and six-membered heterocycles that contain one oxygen, nitrogen or sulfur atom are included in Box 1.1. The ring atoms are normally numbered such that the heteroatom carries the lowest number.

Some authors use Greek letters, α, β and γ, *etc.*, in place of numbers, to indicate the position of substitution in much the same way that the terms *ortho*, *meta* and *para* are used for benzenes.

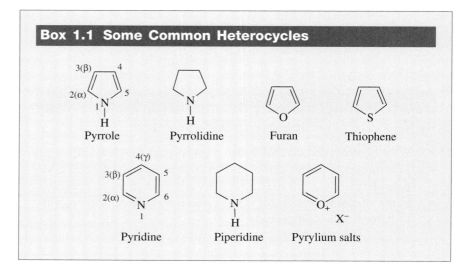

Box 1.1 Some Common Heterocycles

Pyrrole   Pyrrolidine   Furan   Thiophene

Pyridine   Piperidine   Pyrylium salts

A problem arises with trivial names when a sp^3 hybridized atom is present in an otherwise unsaturated ring. A good example is **pyran**, a heterocycle that is formally the product of the addition of a single hydride ion to the pyrylium cation. However, as this addition could occur either at C-2 or C-4, two isomers of pyran are possible; so the question is, how can you distinguish between them? The solution is to call one compound 2*H*-pyran and the other 4*H*-pyran, using the number of the ring atom and the letter H, in italics, to show the position of the hydrogen (see Box 1.2). This system of nomenclature works tolerably well in many related cases and is widely used; other examples will be found in this book.

It is also customary to use the prefixes di-, tetra-, hexahydro- ... (rather than tri-, penta- or heptahydro- ...) when referring to compounds that are partly or fully reduced. This terminology reflects the fact that hydrogen atoms are added two at a time during the hydrogenation of multiple bonds, and it is used even when the compound contains an odd number of hydrogen atoms relative to its fully unsaturated parent. As before, the position of the 'extra hydrogen' atom is located by means of the ring atom number, followed by the letter *H*. It is important to note

Box 1.2 Pyrans

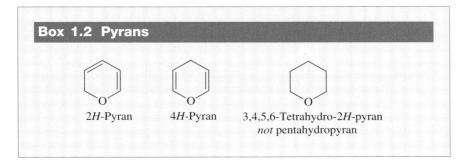

2*H*-Pyran   4*H*-Pyran   3,4,5,6-Tetrahydro-2*H*-pyran
not pentahydropyran

that the *lowest possible number* is always selected for the locant; so, for example, the fully reduced pyrylium cation is referred to as 3,4,5,6-tetrahydro-2*H*-pyran (see Box 1.2).

Since trivial names are so well established, it is now very difficult to abandon them in favour of a logical nomenclature system that provides structural information. Nevertheless, a predictive method of this type is very desirable, especially for molecules where there may be two or more heteroatoms present in the ring. One approach is to use a prefix, which is indicative of the heteroatom [aza (N), oxa (O), thia (S), bora (B), phospha (P), sila (Si), *etc*.], to the name of the corresponding carbocycle. Thus, **pyridine** becomes azabenzene and **piperidine** is azacyclohexane.

This method is useful when dealing with simple heterocycles, but it can become clumsy with more complex ones. An alternative is the **Hantzsch–Widman system**, which uses the same prefixes, but adds a stem name designed to indicate not only the ring size but also the state of unsaturation or saturation (note: when the stem name begins with a vowel the last letter, a, of the prefix is dropped). The stem names for rings containing up to 10 atoms are shown in Table 1.1.

Using this terminology, **furan** becomes oxole and **tetrahydrofuran** is named oxolane; pyridine is azine and piperidine is azinane. As with trivial names, the potential difficulty over partly reduced heterocycles is resolved

Table 1.1 Hantzsch–Widman stem names for heterocycles with 3–10 ring atoms

Ring size	Unsaturated	Saturated
3	irene	irane
4	ete	etane
5	ole	olane
6	ine	inane
7	epine	epane
8	ocine	ocane
9	onine	onane
10	ecine	ecane

For a full discussion of how to name heterocycles by this and other methods, see Panico *et al.*[7]

by using the usual numbered *H* prefix; thus, the four possible isomers of **azepine** are termed as in Box 1.3.[7]

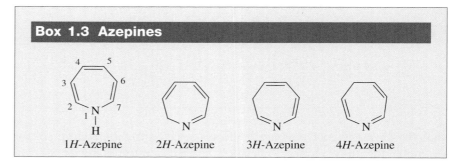

Box 1.3 Azepines

1*H*-Azepine   2*H*-Azepine   3*H*-Azepine   4*H*-Azepine

Many heterocycles are fused to other ring systems, notably benzene, giving in this case **benzo** derivatives; some of these compounds are also extremely well known and have trivial names of their own, such as **indole** and **isoquinoline**. Here, however, it is possible to relate these compounds back to the parent monocycles by indicating to which face the ring fusion applies. To do this, each face of the ring is given a letter (lower case italic), beginning with the face that bears the heteroatom (see Box 1.4).

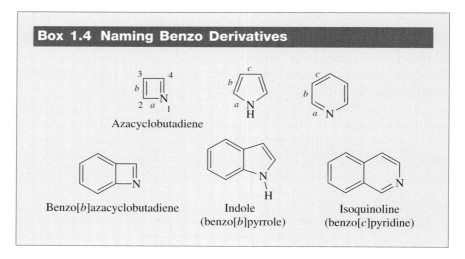

Box 1.4 Naming Benzo Derivatives

Azacyclobutadiene

Benzo[*b*]azacyclobutadiene   Indole
(benzo[*b*]pyrrole)   Isoquinoline
(benzo[*c*]pyridine)

1.3   Importance to Life and Industry

Many heterocyclic compounds are **biosynthesized** by plants and animals and are biologically active. Over millions of years these organisms have been under intense evolutionary pressure, and their metabolites may be used to advantage; for example, as toxins to ward off predators, or as colouring agents to attract mates or pollinating insects. Some heterocycles are fundamental to life, such as **haem** derivatives in blood and the chlorophylls essential for photosynthesis (Box 1.5). Similarly, the paired bases found in **RNA** and **DNA** are heterocycles, as are the sugars that in combination with phosphates provide the backbones and determine the topology of these nucleic acids.

Box 1.5 Some Heterocycles Important for Life

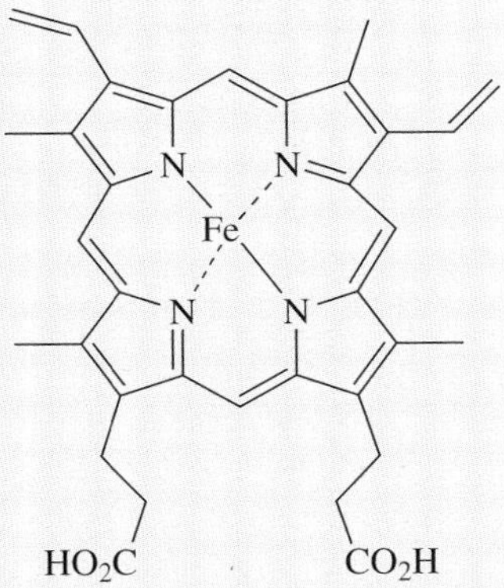

Haem

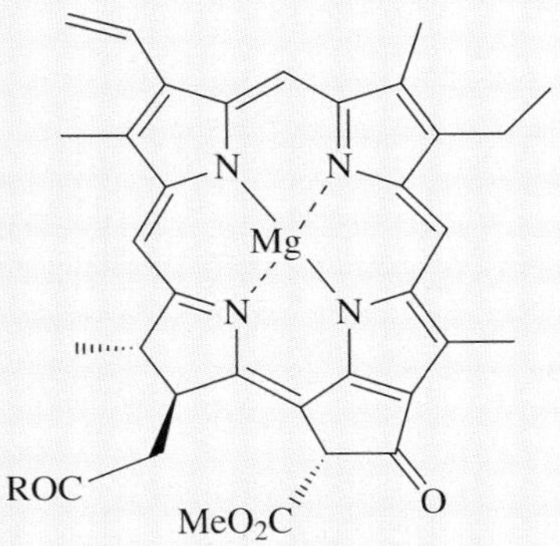

Chlorophyll a

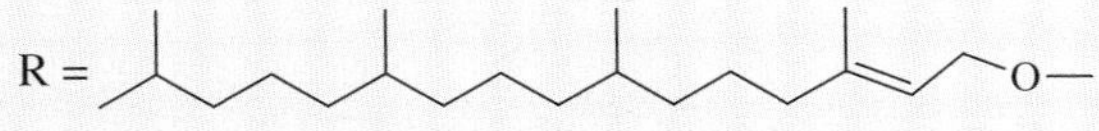

$R =$

Dyestuffs of plant origin include **indigo blue**, used to dye jeans. A poison of detective novel fame is **strychnine**, obtained from the plant resin curare (Box 1.6).

Box 1.6 Some Other Natural Heterocycles

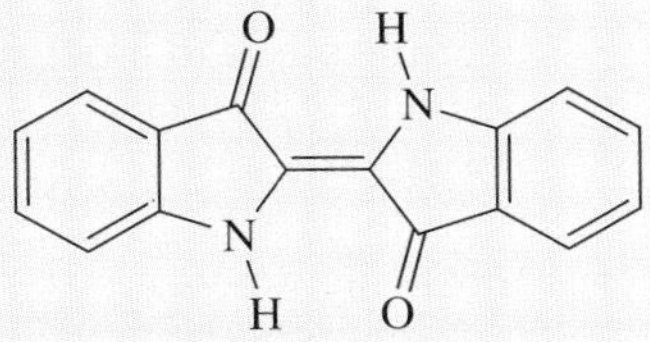

Indigo (indigotin)

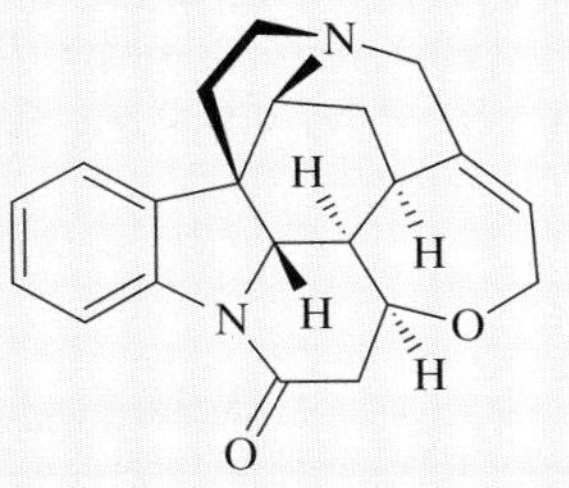

Strychnine

The biological properties of heterocycles in general make them one of the prime interests of the pharmaceutical and biotechnology industries. A selection of just six biologically active pyridine or piperidine derivatives is shown in Box 1.7. It includes four natural products (**nicotine**, **pyridoxine**, **cocaine** and **morphine**) and two synthetic compounds (**nifedipine** and **paraquat**).

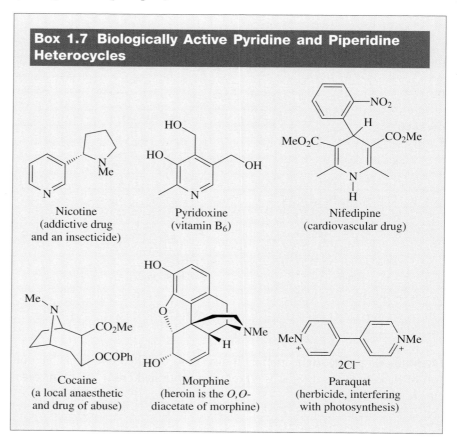

Box 1.7 Biologically Active Pyridine and Piperidine Heterocycles

Nicotine
(addictive drug
and an insecticide)

Pyridoxine
(vitamin B_6)

Nifedipine
(cardiovascular drug)

Cocaine
(a local anaesthetic
and drug of abuse)

Morphine
(heroin is the O,O-
diacetate of morphine)

Paraquat
(herbicide, interfering
with photosynthesis)

There are many thousands of other heterocyclic compounds, both natural and synthetic, of major importance, not only in medicine but also in most other activities known to man. Small wonder then that their chemistry forms a major part of both undergraduate and postgraduate curricula.

1.4   General Principles

1.4.1 Aromaticity

Many fully unsaturated heterocyclic compounds are described as **aromatic**, and some have a close similarity to benzene and its derivatives. For example, pyridine (azabenzene) is formally derived from benzene through the replacement of one CH unit by N. As a result, the constitutions of the two molecules are closely related: in each molecule all the ring atoms are sp^2 hybridized, and the remaining singly occupied p-orbital is orientated at right angles to the plane of the ring (orthogonal). All six p-orbitals overlap to form a delocalized π-system, which extends as a closed loop above and below the ring.

Pyridine and benzene conform to **Hückel's rule**, which predicts that planar cyclic polyenes containing ($4n + 2$) π-electrons ($n = 0$, or an integer) should show added stability over that anticipated for theoretical polyenes composed of formal alternate single and double bonds. This difference is sometimes called the **empirical resonance energy**. For example, benzene, where $n = 1$, is estimated to be 150 kJ mol^{-1} more stable than the hypothetical molecule cyclohexatriene (Box 1.8); for pyridine, the empirical resonance energy is 107 kJ mol^{-1}.

Values for resonance energy can be obtained in several ways, and when comparisons are being made between one molecule and another the data must be obtained by the *same method of calculation*.

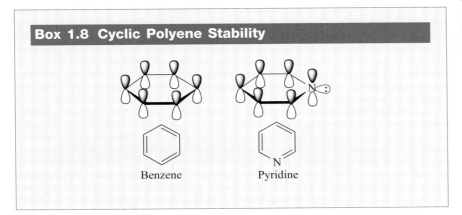

Box 1.8 Cyclic Polyene Stability

Benzene   Pyridine

Alternate double and single bonds are often used in drawing aromatic structures, although it is fully understood these form a closed loop (π-system) of electrons. The reason is that these classical structures are used in the valence bond approach to molecular structure (as canonical forms), and they also permit the use of curly arrows to illustrate the course of reactions.

The increased stability of $4n + 2$ cyclic planar polyenes, relative to their imaginary classical counterparts, comes about because all the bonding energy levels within the π-system are *completely filled*. For benzene and pyridine there are three such levels, each containing two spin-paired electrons. There is then an analogy between the electronic constitutions of these molecules and atoms that achieve noble gas structure.

A further result of the delocalization of the p-electrons is the merging of single and double bonds; benzene is a perfect hexagon with all C–C bond lengths the same (0.140 nm).

Like benzene, pyridine is hexagonal in shape, but in this case the perfect symmetry of the former molecule is distorted because the C–N bonds

If cyclohexatriene were to exist in a localized form and was a planar molecule it would contain three long single bonds and three short double bonds (in buta-1,3-diene the C_1–C_2 bond length is 0.134 nm and the C_2–C_3 bond length is 0.148 nm). The result would be an irregular hexagon and there would be two isomers for, say, a hypothetical 1,2-dichlorocyclo-hexatriene: one with a single C–C bond separating the two chlorine atoms, and the other with a double C=C bond.

are slightly shorter than the C–C bonds (0.134 nm *versus* 0.139–0.140 nm). This is because nitrogen is more electronegative than carbon, and this fact also affects the nature of the π-system. In pyridine the electron density is no longer uniformly distributed around the ring and is concentrated at the N atom.

Another difference between the molecules is that whereas in benzene each carbon is bonded to a hydrogen atom, in pyridine the nitrogen possesses a lone (unshared) pair of electrons. This lone pair occupies an sp^2 orbital and is orientated in the same plane as the ring; moreover, it is available to capture a proton so that pyridine is a base.

In five-membered heterocycles, formally derived from benzene by the replacement of a CH=CH unit by a heteroatom, **aromaticity** is achieved by sharing four p-electrons, one from each ring carbon, with two electrons from the heteroatom. Thus in **pyrrole**, where the heteroatom is N, all the ring atoms are sp^2 hybridized, and one sp^2 orbital on each is bonded to hydrogen. To complete the six π-electron system the non-hybridized p-orbital of N contributes *two* electrons (Box 1.9). It follows that the nitrogen atom of pyrrole no longer possesses a lone pair of electrons, and the compound cannot function as a base without losing its aromatic character.

Box 1.9 Pyrrole

Pyrrole

1.4.2 Non-aromaticity and Anti-aromaticity

Cyclic polyenes and their heterocyclic counterparts which contain $4n$ p-electrons do not show aromaticity, since should these molecules be *forced to form a planar array* the orbitals used to accommodate the electrons within the closed loop are no longer just bonding in nature, but a mixture of *both* bonding and non-bonding types. For a fully unsaturated planar polyene containing four ring atoms, the number of bonding energy levels is one and there are two degenerate non-bonding levels (Box 1.10); in the case of an eight-membered ring, there are three bonding sub-levels and two degenerate non-bonding levels.

Consider a fully delocalized symmetrical 'cyclobutadiene'; here each carbon atom is equivalent and sp^2 hybridized; this leaves four p-electrons to overlap and to form a π-system. Two electrons would then

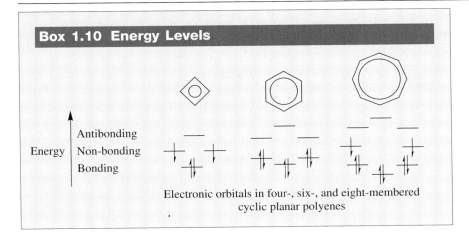

Box 1.10 Energy Levels

Energy

Antibonding
Non-bonding
Bonding

Electronic orbitals in four-, six-, and eight-membered
cyclic planar polyenes

enter the bonding orbital with their spins paired; however, following
Hund's rule the other two have to occupy the two degenerate non-
bonding orbitals singly with their spins parallel. In essence the result is
a **triplet diradical**, which is **anti-aromatic**, *i.e.* the result of delocalization
actually leads to a destabilization of the molecule relative to an alternative model with double and single bonds.

It turns out that cyclobutadiene is not a perfect square (two bonds
are longer than the others), but it is essentially planar. Not surprisingly, it is very unstable and dimerizes extremely readily. It only exists at
very low temperatures either in a matrix with an inert 'solvent' (where
the molecules are kept apart), or at room temperature as an inclusion
compound in a suitable host molecule. Azacyclobutadiene (**azete**) is also
extremely unstable, for similar reasons.

Although a major divergence from planarity is not possible for small
cyclic delocalized polyenes containing $4n$ electrons, their larger equivalents adopt non-planar conformations. Here destabilizing orbital overlap between adjacent double bonds is minimized; the compounds are thus
non-aromatic, and their chemistry often resembles that of a cycloalkene.

A good example is cyclooctatetraene (Box 1.11); formally the higher
homologue of benzene, it is a $4n$ type containing eight p-electrons. This

Hund's rule states: electrons
enter degenerate orbitals singly
with their spins parallel, before
pairing takes place. The term
degenerate here means having
the same energy but not the
same symmetry or spatial
orientation.

The term **triplet** derives from the
three spin states used by a
molecule having two unpaired
electrons. A singlet state is one in
which all the electrons are spin-
paired, and in principle for every
triplet state there is a
corresponding singlet state. In
most cases the triplet state is
more stable than the singlet (also
a consequence of Hund's rule).

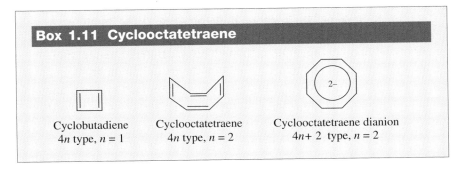

Box 1.11 Cyclooctatetraene

Cyclobutadiene
$4n$ type, $n = 1$

Cyclooctatetraene
$4n$ type, $n = 2$

Cyclooctatetraene dianion
$4n + 2$ type, $n = 2$

The dianion of cyclooctatetraene is planar and aromatic in nature. It has two more electrons than its parent and consequently has 10 π-electrons; it now becomes a member of the aromatic $4n + 2$ series.

The absolute frequency of an ^{1}H NMR signal is not normally measured; instead, tetramethylsilane [$(CH_3)_4$Si, TMS] is added to the sample as an internal standard. The difference between the proton resonance of TMS and that of the sample, both measured in hertz, divided by the spectrometer frequency in megahertz, is called the **chemical shift** (given the symbol δ). This is quoted in ppm (parts per million). To simplify matters the chemical shift of TMS is defined as zero. Note: the vast majority of proton resonances occur downfield from that of TMS, with values greater than 0 ppm.

compound is not planar, it has no special stability and it exists as equilibrating tub-shaped forms with single and double bond lengths of 0.146 nm and 0.133 nm, respectively.

The circulating electrons in the π-system of aromatic hydrocarbons and heterocycles generate a ring current and this in turn affects the **chemical shifts** of protons bonded to the periphery of the ring. This shift is usually greater (downfield from TMS) than that expected for the proton resonances of alkenes; thus ^{1}H NMR spectroscopy can be used as a 'test for aromaticity'. The chemical shift for the proton resonance of benzene is 7.2 ppm, whereas that of the C-1 proton of cyclohexene is 5.7 ppm, and the resonances of the protons of pyridine and pyrrole exhibit the chemical shifts shown in Box 1.12.

Box 1.12 Chemical Shifts

Chemical shifts for the C–H proton resonances of benzene, pyridine and pyrrol e (spectra recorded in CDCl$_3$)

1.4.3 Ring Strain in Cycloalkanes and their Heterocyclic Counterparts

Conformation

Although cyclopropane is necessarily planar, this is not the case for other cycloalkanes. Cycloalkanes utilize sp^3 hybridized carbon atoms, and the preferred shape of the molecule is *partly* determined by the tetrahedral configuration of the bonds. Indeed, any deviation from this ideal induces angle strain. However, other factors must also be considered; for example, although *both* the chair and boat forms of cyclohexane minimize angle strain, the chair form is more stable than the boat by approximately 30 kJ mol^{-1}. This comes about because in the boat representation there are serious non-bonded interactions, particularly C–H bond eclipsing (Box 1.13), that adds to the **torsional strain** of the ring. As a result, only the chair form is populated at normal temperatures. Fully reduced pyridine (piperidine) follows the same pattern and also exists as a chair. However, in this case ring inversion *and* pyramidal inversion of the nitrogen substituents is possible (Scheme 1.1).

Formerly, there was much discussion over how much space a lone pair of electrons occupies relative to a hydrogen atom. It now seems clear

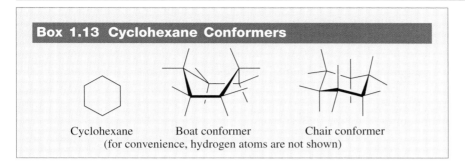

Box 1.13 Cyclohexane Conformers

Cyclohexane   Boat conformer   Chair conformer
(for convenience, hydrogen atoms are not shown)

Scheme 1.1

that there is a preference for an equatorial N–H (*i.e.* H is larger than the lone pair), and this preference is consolidated as the size of the N substituent increases.

The energy difference between equatorial N–H and axial N–H in piperidine is estimated to be 1.5–3.1 kJ mol^{-1} in favour of the equatorial form. In piperidine the energy for N inversion is *ca.* 25.5 kJ mol^{-1}.

Components of Ring Strain

Angle and torsional strain are major components of the total ring strain in fully reduced cyclic compounds. For cycloalkanes (see Table 1.2), the smaller the ring, the larger the overall strain becomes. What may appear at first to be surprising is that medium-sized rings containing 8–11 atoms

Table 1.2 Ring strain in cycloalkanes[8]

Number of atoms in the ring	Total strain (kJ mol^{-1})	Number of atoms in the ring	Total strain (kJ mol^{-1})
3	115	10	52
4	110	11	47
5	26	12	17
6	0.5	13	21.5
7	26	14	8
8	41	15	8
9	53		

are considerably more strained than cyclohexane. One might think that increased flexibility would be beneficial, but in these cases, although puckering reduces angle strain, many pairs of eclipsed H atoms are also created in adjacent CH_2 groups. These may further interact across the ring, causing compression if they encroach within the normal **van der Waals' radii** of the atoms involved (this additional strain is called '**transannular strain**'). However, as more atoms are introduced and the ring size expands, these problems are reduced, and the molecules eventually become essentially strain free.

These considerations may also apply to fully reduced heterocycles, where one or more N or O atoms replace ring carbons, but it must be noted that a change in element also means a change in electronegativity *and* a change of bond length. Thus in hetero analogues of cyclohexane, for example, as C–N and C–O bonds are shorter than C–C bonds, there are increased 1,3- (flagpole) interactions in the chair forms, rendering axial substitution even less favourable.

Furthermore, for multiple replacements, lone pair electrons on the heteroatoms may interact unfavourably and limit certain conformations. In fact, interactions between lone pairs are the main reason for increased barriers to rotation, particularly in N–N bonds compared to C–C single bonds.

Anomeric Effect

When a ring system contains an O–CH–Y unit, where Y is an electronegative group (halogen, OH, OR′, OCOR′, SR′, OR′ or NR′R″), one of the oxygen lone pairs may adopt a *trans* antiperiplanar relationship with respect to the C–Y bond (Box 1.14). In this orientation the orbital containing the lone pair overlaps with the antibonding σ orbital (σ*) of the C–Y bond and 'mixes in' to form a pseudo π-bond. This is called the **anomeric effect**. When Y is F or Cl (strongly electronegative)

Box 1.14 Anomeric Effect

antiperiplanar alignment of one oxygen lone pair and the C–Y bond allows overlap between the lone pair and the empty σ* orbital the net result is that the O–C bond is strengthened and shortened, whereas the C–Y bond is weakened and lengthened. However, for other Y atoms (*e.g.* oxygen or nitrogen) the anomeric effect can operate in both directions, *i.e.* Y can be a donor as well as an acceptor.

Anomeric effects are cumulative, and can cause a potentially flexible ring to adjust to a more rigid conformation in order to maximize the overlap of suitable lone pair and σ* orbitals. It has been particularly instructive in explaining 'anomalous' preferences for substituent orientations in tetrahydropyrans and related compounds. In the case of 2-methoxytetrahydropyran, for example, the axial conformer is three times more populated than the equatorial form (Scheme 1.2).

The **anomeric effect** is not simply restricted to ring compounds and a full discussion of the topic is complex. In general, 'there is a preference for conformations where the best donor lone pair, or bond, is orientated antiperiplanar to the best acceptor bond'.[9]

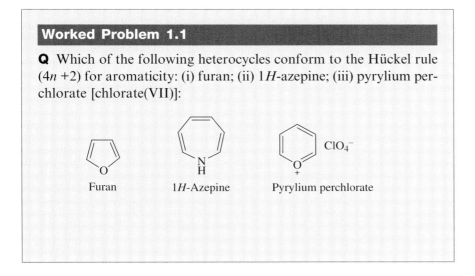

axial (75%)   equatorial (25%)
2-Methoxytetrahydropyran (Y = OMe)

Scheme 1.2

Heteroatom Replacement

Nitrogen and oxygen are found in level 2 of the Periodic Table, and a further alteration in ring topology may arise when the heteroatom is replaced by an element from a lower level. Here, apart from an increase in atomic diameter, the replacement element may use a hybridization state different than that of the earlier elements. Not only can this affect the shape of the molecule, it can also modify the chemical properties.

Worked Problem 1.1

Q Which of the following heterocycles conform to the Hückel rule ($4n + 2$) for aromaticity: (i) furan; (ii) 1*H*-azepine; (iii) pyrylium perchlorate [chlorate(VII)]:

Furan   1*H*-Azepine   Pyrylium perchlorate

A The answer to this question is based upon assuming *at first* the ring to be planar, then counting the number of all the electrons that could contribute to a delocalized π-system. For planar aromatic compounds the number should conform to $4n + 2$. If it does not then the ring is either non-planar or anti-aromatic!

(i) Electronically, furan resembles pyrrole, utilizing four p-electrons from the buta-1,3-diene (C_4) component and one lone pair from oxygen, giving six in all. The molecule is planar and aromatic in character.

(ii) 1*H*-Azepine may well contain six p-electrons, associated with the six carbon atoms of the ring, but an aromatic system should be planar. Were this to be the case, then the lone pair electrons on the nitrogen atom would also overlap with this delocalized system, so that in total there would be eight electrons. Planar 'azepine' would then be a member of the $4n$ ($n = 2$) class and anti-aromatic. In fact, 1*H*-azepine is very difficult to isolate, but stable derivatives are known and have been shown to be non-planar.

(iii) In the *classical* Kekulé representation shown the oxygen atom of the pyrylium cation is trivalent and carries a positive charge. However, the heteroatom can still contribute one p-electron to a sextet of π-electrons, five of which are supplied by the five ring carbon atoms. Pyrylium salts thus comply with the Hückel rule, but we shall see later (Chapter 4) that the electronegative oxygen strongly influences their behaviour.

Worked Problem 1.2

Q Deduce the preferred conformations of (i) 1-*tert*-butylpiperidine [1-(2-methylprop-2-yl)piperidine] and (ii) *trans*-2-methoxy-4-methyltetrahydropyran:

1-*tert*-Butylpiperidine   *trans*-2-Methoxy-4-methyltetrahydropyran

A (i) The *tert*-butyl group is sufficiently bulky that it can only be accommodated in an equatorial site in piperidine. As a result, the ring is locked in a single chair conformation:

Equatorial 1-*tert*-butylpiperidine

(ii) If the methyl group of 2-methoxy-4-methyltetrahydropyran resides in an equatorial site (of course, it is larger than hydrogen!), it then follows that the *trans* methoxy group at C-2 is axially orientated. In this arrangement there are also reinforcing anomeric interactions involving a lone pair from each oxygen atom. Consequently, this conformation is favoured, by a ratio of 98:2, over the alternative in which the methyl group is axial and the methoxy group is equatorial:

trans-2-Methoxy-4-methyltetrahydropyran

Summary of Key Points

1. Planar cyclic polyenes containing $(4n + 2)$ π-electrons obey Hückel's rule for aromaticity and show greater stability than that predicted from their classical structures.

2. The replacement of a CH group by an atom, such as N, O or S, also leads to aromatic heterocycles.

3. Although the conformations of heterocycles are governed by the same principles that apply to carbocycles, where appropriate, additional factors, such as the anomeric effect, can have a significant influence upon the energies of the isomers in equilibrium.

Problems

1. Suggest names for the compounds (a)–(f) shown below:

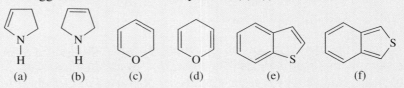

(a) (b) (c) (d) (e) (f)

2. Which of the following compounds (a)–(e) are aromatic and which are non-aromatic or anti-aromatic? Give your reasons.

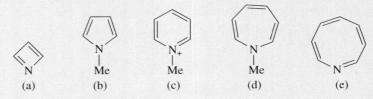

(a) (b) (c) (d) (e)

3. Assuming there are no solvent effects, which isomer is likely to predominate in an equilibrium between the conformers **A** and **B**?

A **B**

References

1. T. L Gilchrist, *Heterocyclic Chemistry*, 2nd edn., Longman/Wiley, Harlow/Chichester, 1992.
2. J. A. Joule and K. Mills, *Heterocyclic Chemistry,* 4th edn., Blackwell Science, Oxford, 2000.
3. A. R. Katritzky, *Handbook of Heterocyclic Chemistry*, Pergamon Press, Oxford, 1985.
4. A. R. Katritzky and C. W. Rees (eds.), *Comprehensive Heterocyclic Chemistry*, vols. 1–8, Pergamon Press, Oxford, 1984.
5. A. R. Katritzky, C. W. Rees and E. F. V. Scriven (eds.), *Comprehensive Heterocyclic Chemistry II, A Review of the Literature 1982–1995*, vols. 1–11, Pergamon Press, Oxford, 1996.
6. *Rodd's Chemistry of Carbon Compounds*, 2nd edn., vols. IVA–K, Elsevier, Amsterdam, 1973–1986 (supplements 1990–2000).
7. R. Panico, W. H. Powell and J.-C. Richer (eds.), *A Guide to IUPAC Nomenclature of Organic Compounds (Recommendations 1993)*, Blackwell Science, Oxford, 1993.

8. J. S. Chickos *et al.*, *J. Org. Chem.*, 1992, **57**, 1897.
9. A. J. Kirby, *The Anomeric Effect and Related Stereoelectronic Effects at Oxygen*, Springer, New York, 1983.

Further Reading

J. Rigaudy and S. P. Klesney (eds.), *IUPAC Nomenclature of Organic Chemistry (Sections A to H)*, Pergamon Press, Oxford, 1979.

L. A. Paquette, *Principles of Modern Heterocyclic Chemistry*, Benjamin, New York, 1966.

A. R. Katritzky, *Physical Methods in Heterocyclic Chemistry*, Academic Press, New York, 1960–1972.

M. J. Cook, A. R. Katritzky and P. Linda, *Aromaticity of Heterocycles*, in *Adv. Heterocycl. Chem.*, 1974, **17**, 257.

D. H. R. Barton and W. D. Ollis (eds.), *Comprehensive Organic Chemistry*, vol. 4, *Heterocyclic Chemistry*, ed. P. G. Sammes, Pergamon Press, Oxford, 1979.

A. R. Katritzky, M. Karelson and N. Malhotra, *Heterocyclic Aromaticity*, in *Heterocycles*, 1991, **32**, 127.

B. Ya. Simkin and V. I. Minkin, *The Concept of Aromaticity in Heterocyclic Chemistry*, in *Adv. Heterocycl. Chem.*, 1993, **56**, 303.

E. L. Eliel and S. H. Wilen, *Stereochemistry of Organic Compounds*, Wiley, Chichester, 1994.

E. Juaristi and G. Cuevas, *The Anomeric Effect*, CRC Press, Boca Raton, Florida, 1995.

2

Pyridine

Aims

By the end of this chapter you should understand:

- The aromaticity and structure of pyridine
- The nature of its reactions with electrophiles, including acids
- The reasons why nucleophilic additions occur readily, particularly in the case of pyridinium salts
- The reactions of some prominent pyridines and piperidines
- Ways in which pyridines and *N*-alkylpiperidines are synthesized. The importance of the Hantzsch synthesis of pyridines

2.1   Resonance Description

A dipole moment results when a molecule has a *permanent* uneven electron density. Individual charge separations in bonds cannot be measured, only the vectorial sum of *all* individual bond moments. A dipole moment is expressed in debye units (D). Only completely symmetrical molecules fail to have a dipole moment, but few have dipole moments greater than 7 D.

As our first more detailed foray into heterocyclic chemistry we will consider pyridine (azabenzene). It is an aromatic compound (see previous chapter), but the replacement of CH by more electronegative N induces a dipole moment of 2.2 D, denoting a shift of electron density from the ring towards the nitrogen atom (benzene, which is symmetrical, has no dipole moment). The valence bond (resonance) description indicates that the nitrogen atom of pyridine carries a partial negative charge and the carbons 2(6) and 4 bear partial positive charges. The canonical forms shown in Scheme 2.1 may then effectively represent the molecule.

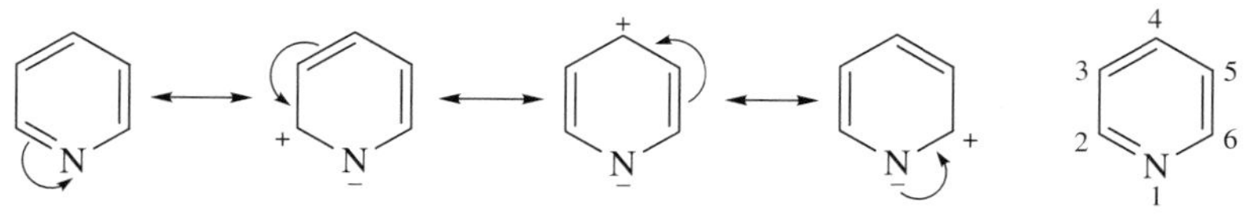

Scheme 2.1

2.2   Electrophilic Substitution

2.2.1 Attack at Nitrogen and at Carbon

From the resonance description you might conclude that although the primary site for electrophilic attack is at N-1, reactions at carbon C-3(5) might be possible, even if not as likely. However, an important point to remember is that the N atom of pyridine carries a lone pair of electrons; these electrons are NOT part of the π-system. As a result, pyridine is a *base* (pK_a 5.2), reacting with acids, Lewis acids and other electrophiles (E⁺) to form stable **pyridinium salts** (Scheme 2.2), in which the heterocycle retains aromatic character.

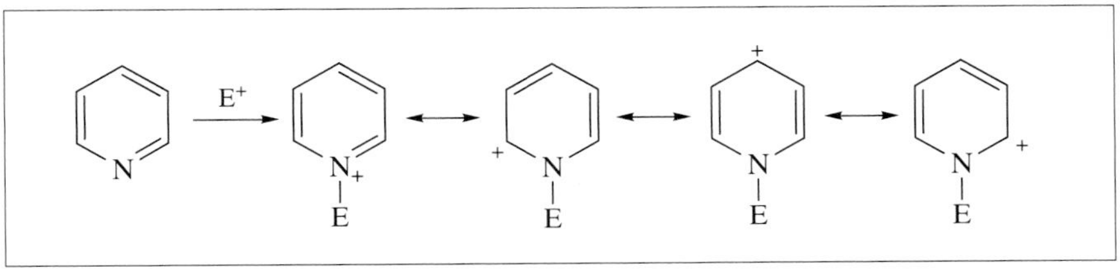

Scheme 2.2

Direct attack at a ring carbon, even C-3, is normally slow (a) because the concentration of free pyridine in equilibrium with the pyridinium salt is extremely low, and (b) attack upon the salt would also require the positive pyridinium cation to bond to a positively charged reactant.

Indeed, where reactions at a ring carbon take place under relatively mild conditions, *special circumstances* are at work. For example, 2,6-*tert*-butylpyridine combines with sulfur trioxide in liquid sulfur dioxide at −10 °C to give the corresponding 3-sulfonic acid (Scheme 2.3). An explanation is that the bulky *tert*-butyl groups prevent access of the 'large' electrophile to N-1. Steric hindrance is much less at C-3 and sulfonation is diverted to this site using the 'free' pyridine as the substrate.

Pyridinesulfonic acids are strongly acidic, so that the 3-sulfonic acid that forms then protonates a second molecule of 2,6-*tert*-butylpyridine (*N*-protonation is permitted because of the small size of the proton). Once protonated, however, further electrophilic attack is strongly disfavoured, and so the overall conversion is limited to 50%.

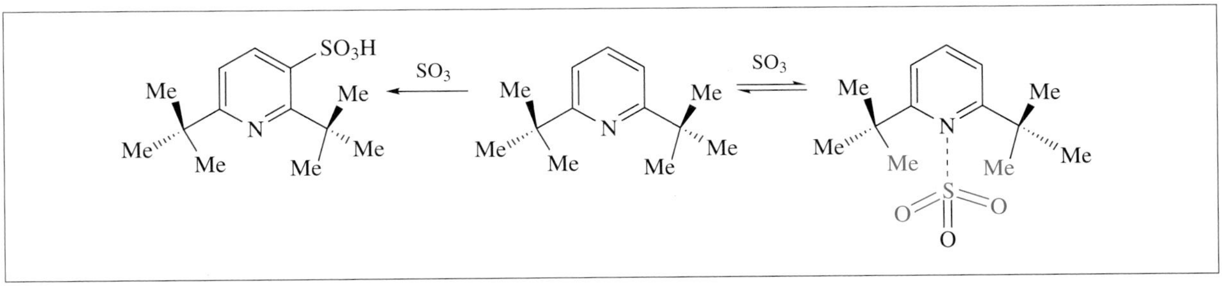

Scheme 2.3

2.2.2 Addition–Elimination

Another feature that is clear from the resonance description of the pyridinium cation is that attack by nucleophiles is favoured at C-2(6) and C-4. This has importance in some reactions where at first sight it may appear that electrophilic reagents combine quite easily with pyridine. These reactions are more subtle in nature!

For example, 3-bromopyridine is formed when pyridine is reacted with bromine in the presence of oleum (sulfur trioxide in conc. sulfuric acid) at 130 °C (Scheme 2.4). Direct electrophilic substitution is *not* involved, however, as **zwitterionic** (dipolar) pyridinium-*N*-sulfonate is the substrate for an *addition* of bromide ion. Subsequently, the dihydropyridine that is formed reacts, possibly as a dienamine, with bromine to generate a dibromide, which then eliminates bromide ion from C-2. It is notable that no bromination occurs under similar conditions when oleum is replaced by conc. sulfuric acid alone; instead, pyridinium hydrogensulfate is produced.

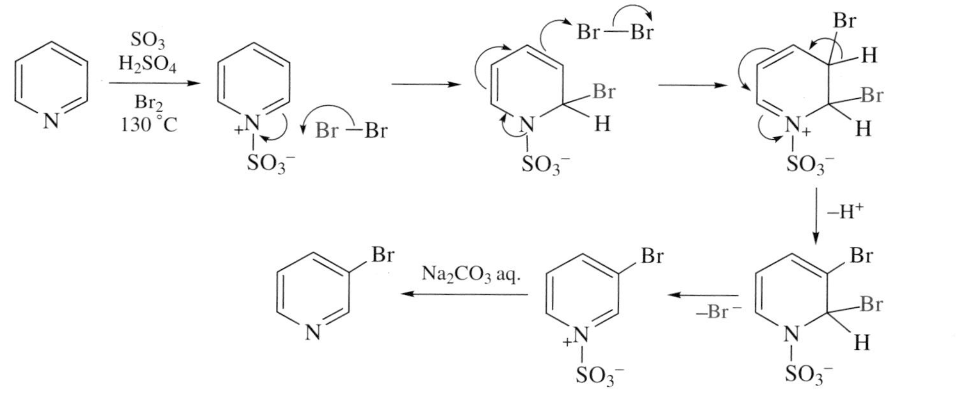

Scheme 2.4

Similarly, pyridine can be 3-sulfonated with hot sulfuric acid, or oleum, if mercuric [mercury(II)] sulfate is present as a catalyst (Scheme 2.5). The process is not straightforward and may involve a *C*-mecuriated pyridine intermediate [it is known, for example, that pyridine reacts with mercuric acetate at room temperature to form a pyridinium salt that decomposes at 180 °C into 3-(acetoxymercuri)pyridine (X = OAc)]. Without the catalyst, long reaction times and a temperature of 350 °C are necessary; even then, the yield of pyridine-3-sulfonic acid is poor.

2.2.3 Acylation and Alkylation

Pyridine reacts with acyl chlorides, or acid anhydrides, to form *N*-acylpyridinium salts, which are readily hydrolysed (Scheme 2.6a).

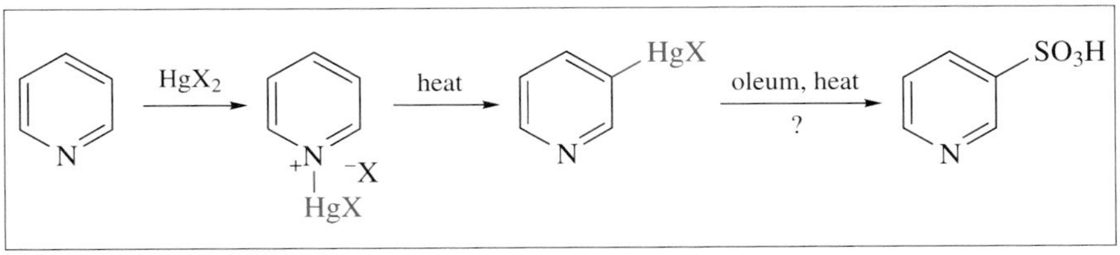

Scheme 2.5

However, the salts can be used as valuable **transacylating agents**, particularly for alcohols, and in this application the salt is not isolated but reacted *in situ* with the alcohol. An excess of pyridine is needed and such reactions were carried out in pyridine both as reagent and as solvent. Unfortunately, pyridine is difficult to remove from the products, and its use has been superseded by **DMAP** [4-(N,N-dimethylamino)pyridine]. Now, after N-acylation the 4-N,N-dimethylamino group reinforces the nucleophilicity of the corresponding acylpyridinium salt (Scheme 2.6b), and this promotes the transfer of the acyl group from the salt to the alcohol in the next step. Only a catalytic amount of DMAP is used.

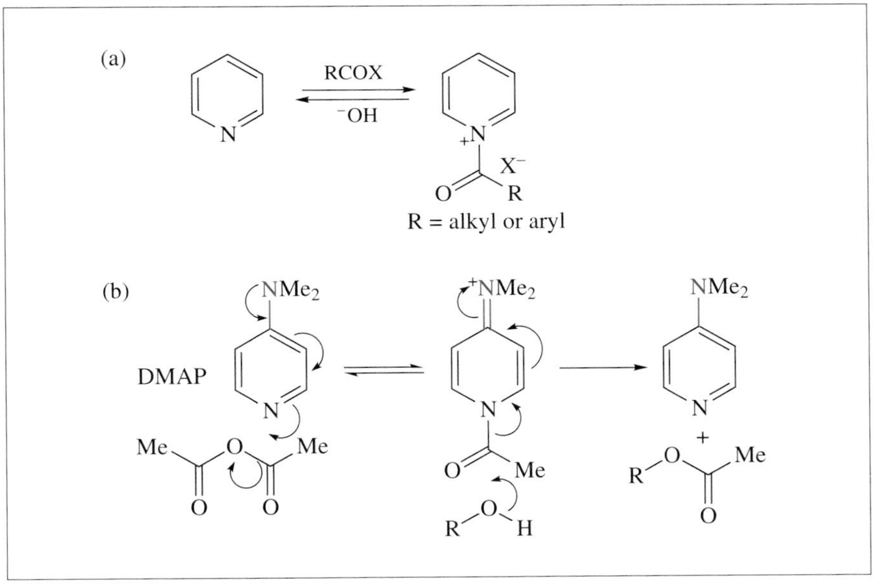

Scheme 2.6

Alkyl halides and related alkylating agents react with pyridines to form *N*-alkylpyridinium salts (Scheme 2.7). These compounds are much more stable than their N-acylpyridinium equivalents and can often be isolated as crystalline solids, particularly if the halide ion is exchanged for perchlorate, tetrafluoroborate or another less polarizable counter anion.

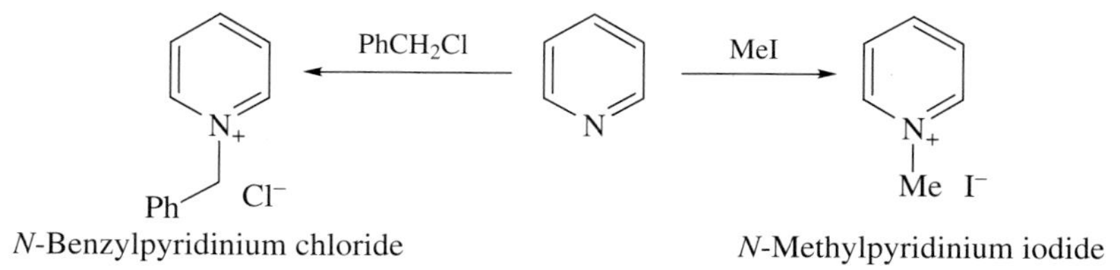

N-Benzylpyridinium chloride   N-Methylpyridinium iodide

Scheme 2.7

2.3   Pyridine *N*-Oxides

Scheme 2.8

Pyridine *N*-oxides are frequently used in place of pyridines to facilitate electrophilic substitution. In such reactions there is a balance between electron withdrawal, caused by the inductive effect of the oxygen atom, and electron release through resonance from the same atom in the opposite direction. Here, the resonance effect is more important, and electrophiles react at C-2(6) and C-4 (the antithesis of the effect of resonance in pyridine itself).

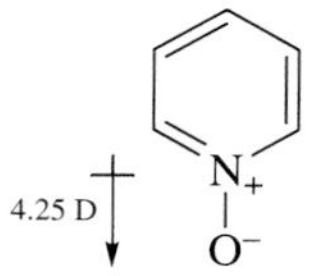

The *N*-oxide is prepared from pyridine by the action of a **peracid** (*e.g.* hydrogen peroxide in acetic acid, forming peracetic acid *in situ*, or *m*-chloroperbenzoic acid, MCPBA); pyridine is regenerated by deoxygenation by heating with triphenylphosphine (Ph$_3$P $\rightarrow$ Ph$_3$PO) (Scheme 2.8).

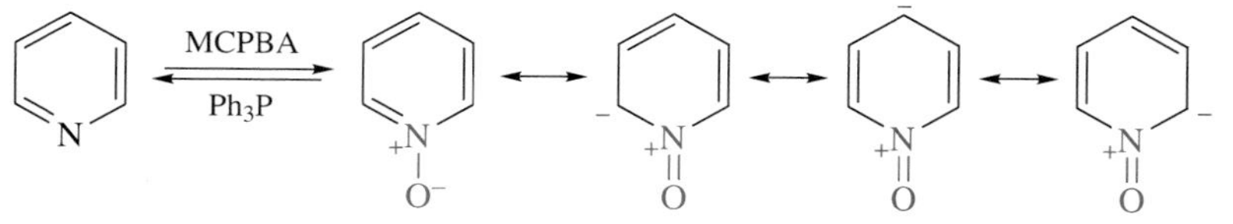

As long as the conditions are selected so that the *N*-oxygen atom is not irreversibly protonated, reactions with electrophiles give 2- and 4-substituted products. Thionyl chloride, for example, gives a mixture of 2- and 4-chloropyridine *N*-oxides in which the 4-isomer is predominant. However, pyridine *N*-oxide reacts with acetic anhydride first to give 1-acetoxypyridinium acetate and then, on heating, to yield 2-acetoxypyridine through an addition–elimination process (Scheme 2.9a). When a similar reaction is carried out upon the 2,3-dimethyl analogue, the acetoxy group rearranges from N-1 to the C-2 methyl group, at 180 °C, to form 2-acetoxymethyl-3-methylpyridine (possibly as shown in Scheme 2.9b).

Nitration at C-4 occurs with conc. sulfuric acid and fuming nitric acid (Scheme 2.10a); very little 2-nitropyridine *N*-oxide is formed, but in cases where the electrophile binds strongly to the oxygen atom of the *N*-oxide, further attack occurs at C-3. Thus, pyridine *N*-oxide is brominated at 70 °C by bromine and oleum to form 3-bromopyridine *N*-oxide, and sulfonated by oleum and mercuric sulfate at 240 °C to give pyridine-3-sulfonic acid *N*-oxide (Scheme 2.10b).

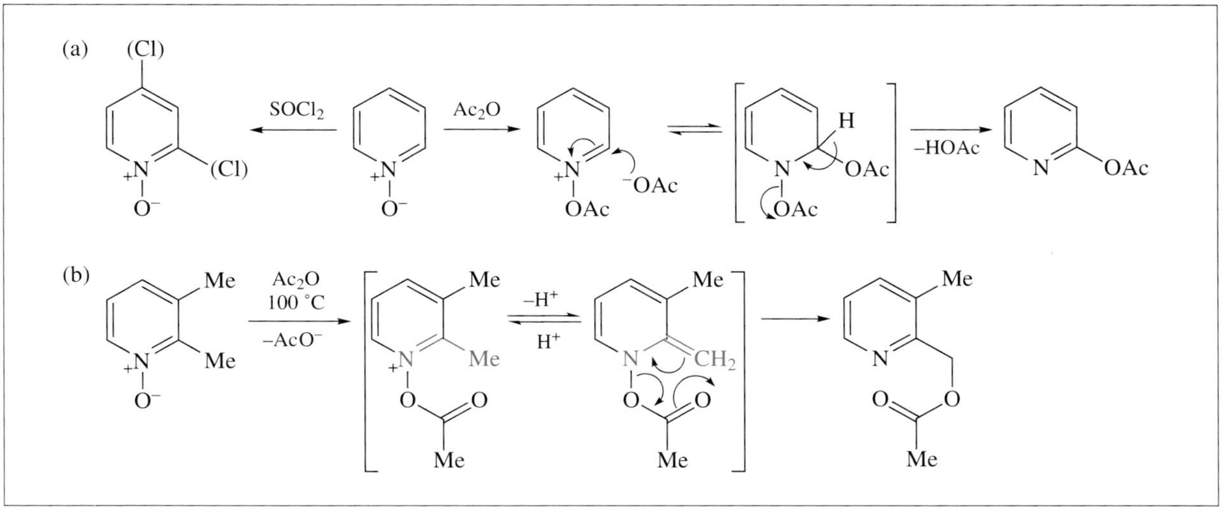

Scheme 2.9

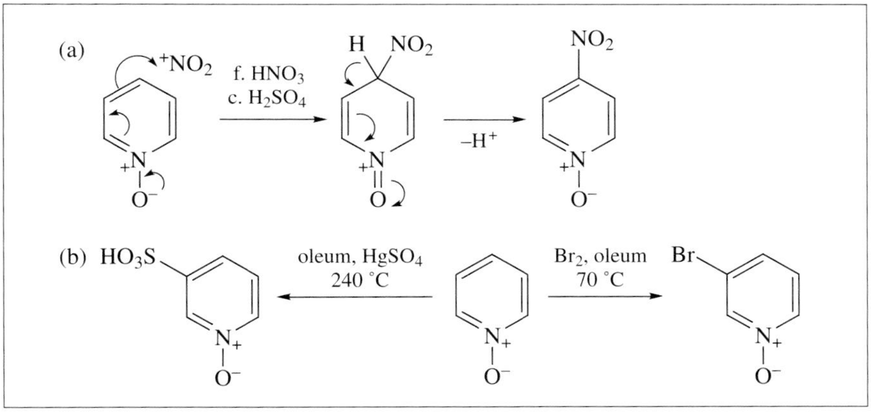

Possibly the substrate for the last two reactions is the *N*-sulfonyloxypyridinium cation:

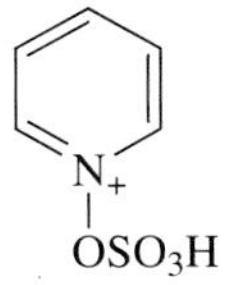

Scheme 2.10

2.4   Nucleophilic Substitution

2.4.1 The Effects of the Pyridine Resonance and Leaving Groups

When pyridine is reacted with nucleophiles the attack occurs preferentially at C-2(6) and/or at C-4, as predicted by the resonance description of possible reaction intermediates (Scheme 2.11). The problem,

An intermediate formed through attack at C-3(5) would *not* permit the negative charge to be shared with the N atom.

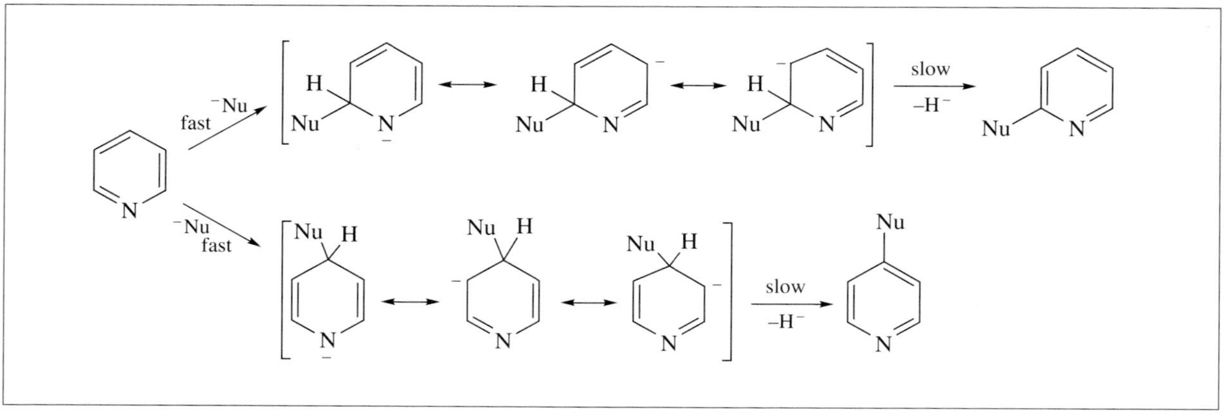

Scheme 2.11

however, is that for unsubstituted pyridines the leaving group is the highly reactive hydride ion. So, although the first step in the reaction is favoured, the second step is not. Oxygen in the air, or an added oxidant, may ease the situation and serve to oxidize the intermediate to an aromatic pyridine.

2.4.2 Chichibabin Reaction

A classic reaction of this type is **Chichibabin amination**, leading mainly to 2-aminopyridine (Scheme 2.12a). This takes place when a pyridine is heated at 140 °C with sodamide (NH_2^- is a very strong nucleophile). Although hydrogen gas is certainly evolved during the reaction, the initial proton donor is not known. However, once some 2-aminopyridine is formed this product could function as the donor (Scheme 2.12b), and the process may then become a form of chain reaction.

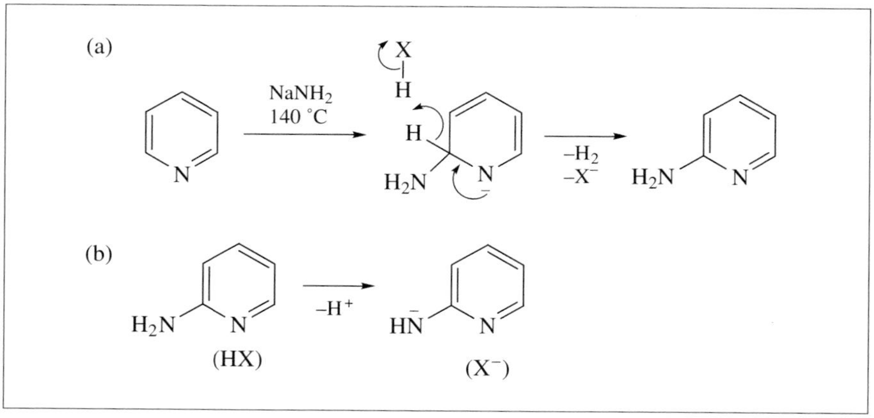

Scheme 2.12

2.4.3 Nucleophilic Reactions of Halopyridines

Halide Ion versus Hydride Ion

Normally, nucleophilic attack occurs preferentially at C-2(6); this selectivity is the result of the enhanced **inductive effect** experienced by the carbon atoms immediately adjacent to the more electronegative nitrogen (Scheme 2.13). If both C-2 and C-6 are occupied, then attack at C-4 takes place. However, it is possible to influence the site and rate of the reaction if a potential leaving group replaces hydrogen. After addition, the loss of the leaving group from the σ-intermediate will be easier than if it were the very reactive hydride ion. **Halopyridines** are often used, although not exclusively, and this normally ensures preferential nucleophilic substitution at the site of the halogen atom.

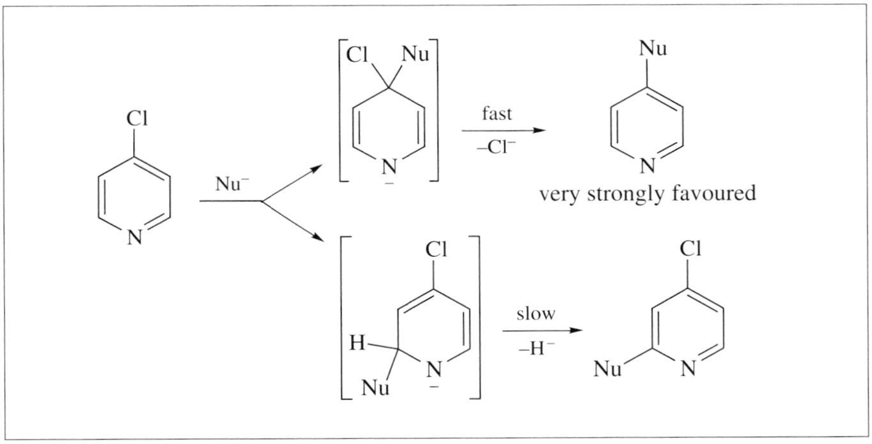

Scheme 2.13

'Addition–substitution' easily occurs with a variety of nucleophilic reagents, including NaOMe, PhSH, PhNHMe and NH_3. Thus, with 2-chloropyridine a range of 2-substituted pyridines is formed (Scheme 2.14).

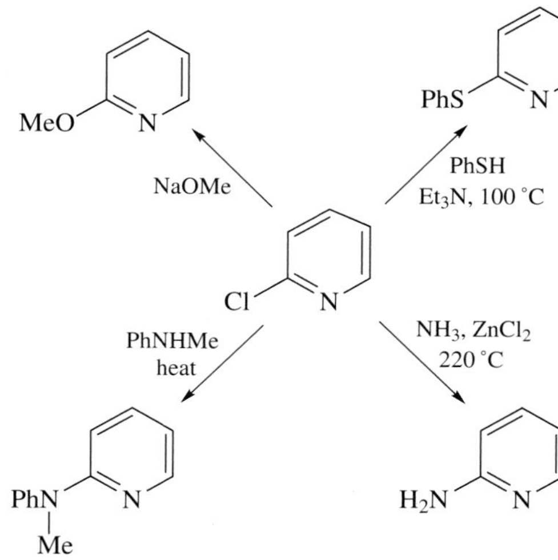

Scheme 2.14

Worked Problem 2.1

Q Outline a synthesis of 2-acetoxy-4-methoxypyridine from pyridine.

A This synthesis requires several steps, and illustrates the way pyridine *N*-oxides can be used in the electrophilic substitution of pyridines. It also indicates selectivity in the reactions of nucleophiles with pyridinium salts. The first step is to convert pyridine into its *N*-oxide, and to react this with conc. nitric acid and conc. sulfuric acid at 160 °C (Scheme 2.15). This provides the 4-nitro derivative in which the nitro group directs the position of attack by sodium methoxide in methanol, by providing a good leaving group (NO$_2^-$) in the next step. The product 4-methoxypyridine *N*-oxide can now be reacted with hot acetic anhydride, which first causes *O*-acetylation and the formation of an intermediate *O*-acetoxypyridinium salt. This adds acetate anion at C-2 and, without isolation, the adduct 1,2-diacetoxy-4-methyl-1,2-diydropyridine which is produced eliminates acetic acid, thus rearomatizing the heterocycle and forming the required pyridine

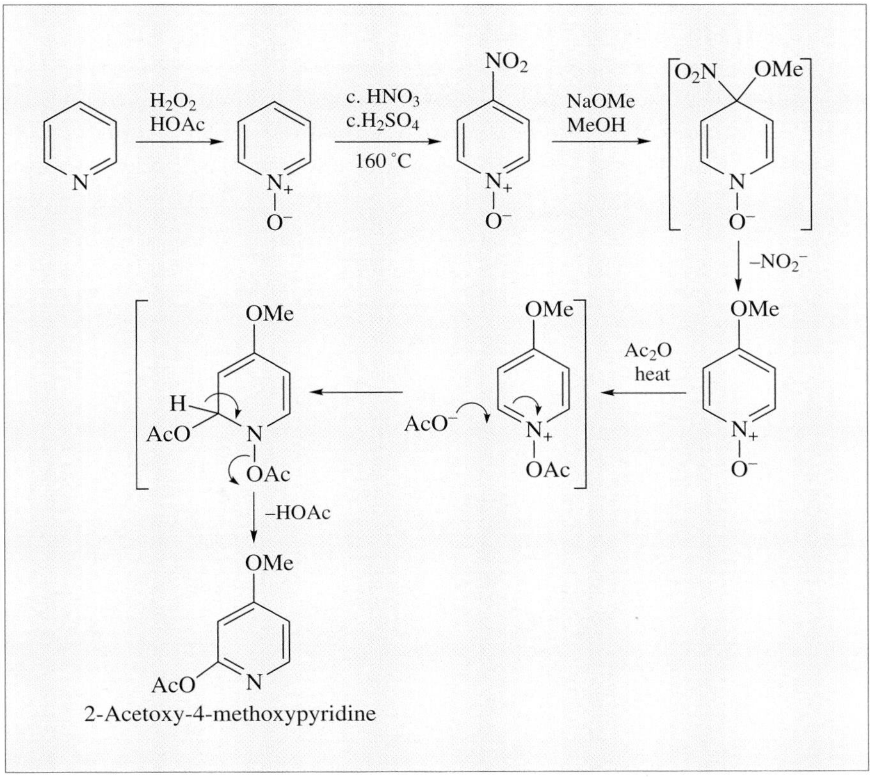

Scheme 2.15

2.4.4 Pyridynes

There is a complication if the nucleophile used in reactions with halopyridines is also a strong base; for now the formation of a **pyridyne** is possible, and with sodamide in liquid ammonia (providing the NH$_2^-$ ion?, B), for example, both 3-aminopyridine and 4-aminopyridine are formed from 4-bromopyridine (Scheme 2.16).

Scheme 2.16

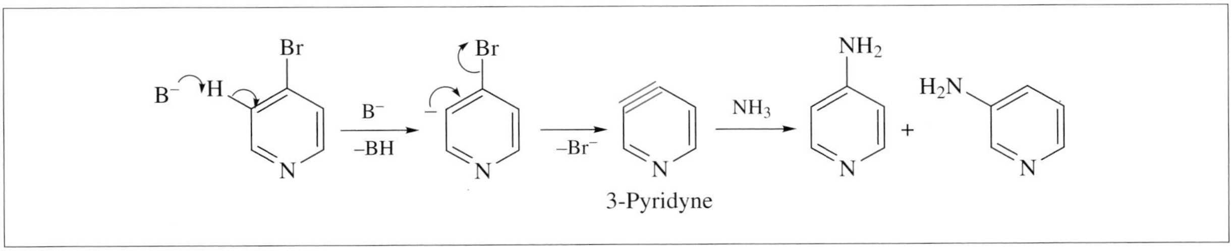

This occurs because 3-pyridyne (3,4-didehydropyridine) is formed by an **E1cB process** [elimination (first order) from the conjugate base]. 3-Pyridyne then adds ammonia; the addition is not regiospecific and *two* amino derivatives are formed.

An alternative mechanism is the addition of NH$_2^-$ to 3-pyridyne, followed by protonation as the second step.

2.5 Lithiation

Halogenopyridines can undergo **metal–halogen exchange** when treated with butyllithium. The lithium derivatives then behave in a similar manner to aryllithiums and Grignard reagents and react with electrophiles such as aldehydes, ketones and nitriles (Scheme 2.17). Thus, aldehydes and ketones form alcohols, and nitriles yield *N*-lithioimines, which on hydrolysis are converted into pyridyl ketones.

Scheme 2.17

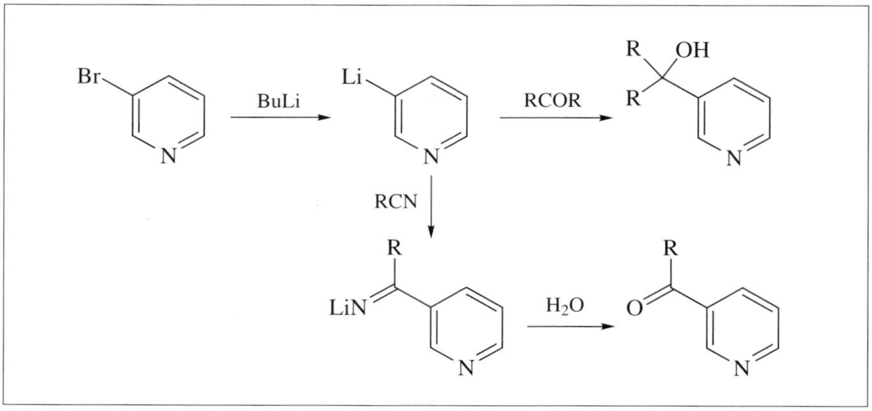

Normally, compounds containing a methine group situated next to a carbonyl group (CH–CO) are in equilibrium with an enol tautomer which contains the C=C(OH) group. The term **enolate anion** refers to the resonance-stabilized anion, formed by deprotonating the enol tautomer with a sufficiently strong base:

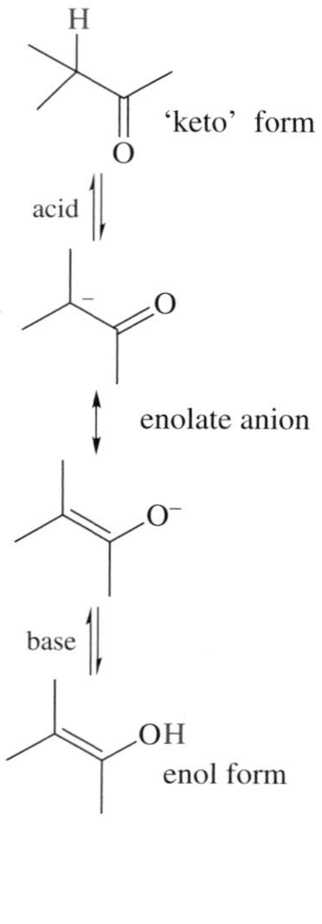

2.6 Methods of Synthesis

2.6.1 Hantzsch Synthesis

The most used route to pyridines is called the **Hantzsch synthesis**. This uses a 1,3-dicarbonyl compound, frequently a 1,3-keto ester [ethyl acetoacetate (ethyl 3-oxobutanoate)], and an aldehyde, which are heated together with ammonia (Scheme 2.18). At the end of the reaction the dihydropyridine is oxidized to the corresponding pyridine with nitric acid (or another oxidant such as MnO_2). The normal Hantzsch procedure leads to symmetrical dihydropyridines. Two different 1,3-dicarbonyl compounds may not be used as two enolate anions might form, giving mixed products when reacted with the aldehyde. The aldehyde itself should preferably be non-enolizable, otherwise the chance of aldolization exists, but with care this can be avoided.

2.6.2 Guareschi Synthesis

This is a similar synthesis in which the ring atoms are assembled by reacting a 1,3-dicarbonyl compound with cyanoacetamide (cyanoethanamide) under mildly basic conditions (Scheme 2.19). The product, a 3-cyano-2-pyridone, may then be hydrolysed and decarboxylated, before the oxygen atom of the carbonyl group is removed in two steps: by chlorination and hydrogenolysis.

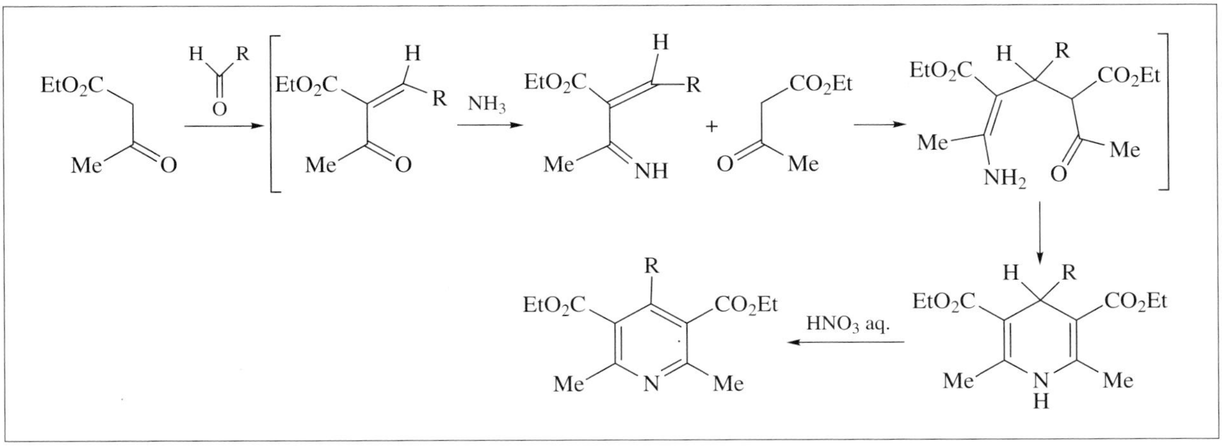

Scheme 2.18

Scheme 2.19

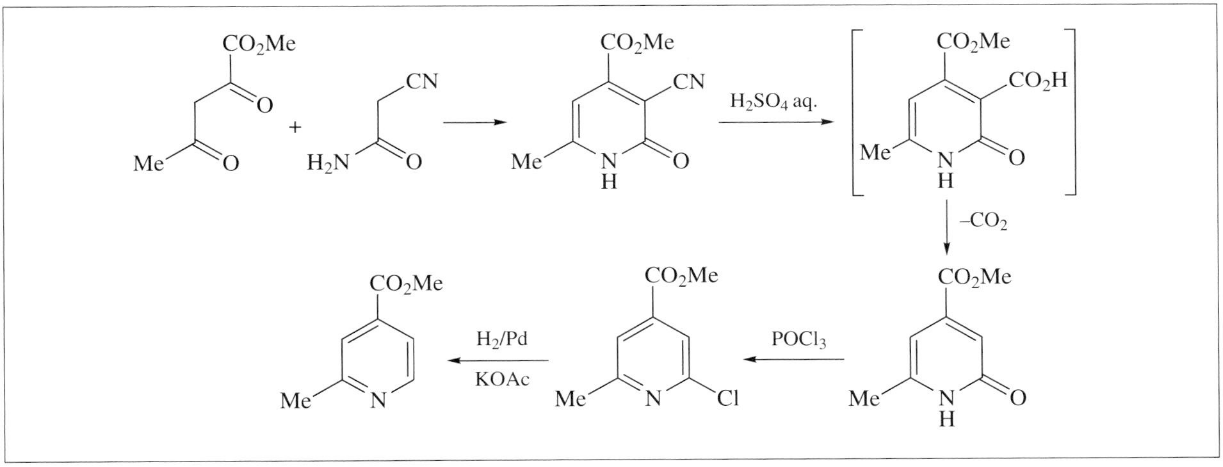

2.7 Commonly Encountered Pyridine Derivatives

The chemical behaviour of many substituents attached to the pyridine ring is similar to that of the corresponding groups in benzene, with the *proviso* that resonance with, and the inductive influence of, the ring nitrogen atom may significantly modify some reactions.

2.7.1 Methylpyridines (Picolines)

These have the trivial generic name picolines; on oxidation they give the appropriate acids; pyridine-2- and -4-carboxylic acids are called 2- and

4-picolinic acids, respectively (see Box 2.1). Pyridine-3-carboxylic acid is better known as **nicotinic acid** (reflecting its relationship to the alkaloid nicotine, see Section 1.3).

Box 2.1 Picolines

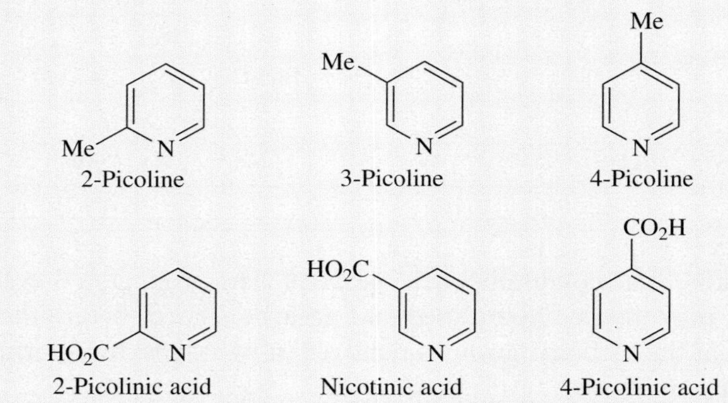

2-Picoline and 4-picoline are easily deprotonated as the conjugate carbanions are resonance stabilized. These anions can be used in alkylations and other reactions with electrophiles (Scheme 2.20).

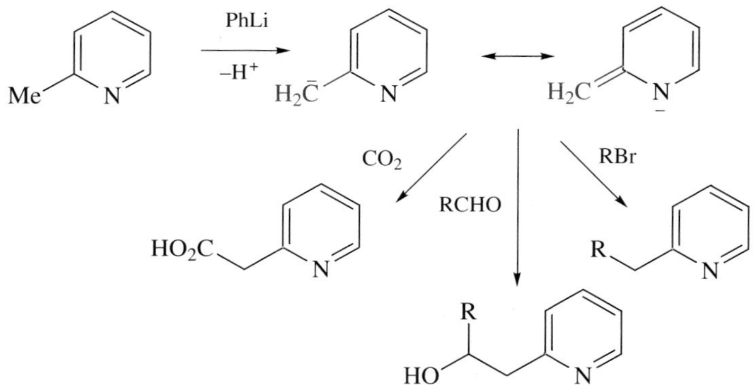

Scheme 2.20

Deprotonation of 3-picoline is more difficult (the anion cannot achieve stability through resonance, as happens with the others) and a much stronger base, LDA [lithium diisopropylamide (lithium propan-2-ylamide)], is needed. Once achieved, however, the conjugate anion behaves as a nucleophile and undergoes typical carbanion reactions (indeed, it is more reactive than its counterparts, since reactivity is most often the opposite of stability!).

Worked Problem 2.2

Q Provide a synthesis of stilbazole (**2.1**) from 2-methylpyridine.

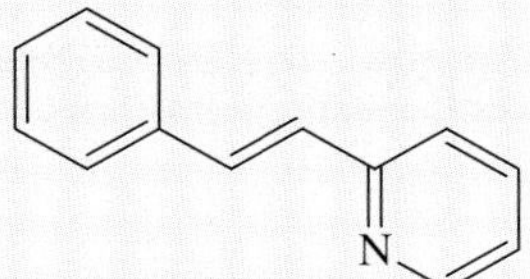

2.1 Stilbazole

A 2-Methylpyridine can be deprotonated by a base such as sodium methoxide and the resultant anion can be reacted with benzaldehyde to form a hydroxylated adduct (Scheme 2.21). This product can then be dehydrated by acid, or base, to form the conjugated compound stilbazole. Alternatively 2-methylpyridine can be *N*-acetylated by reaction with acetic anhydride and the initial product deprotonated to give an *N*-acetylenamine intermediate that traps benzaldehyde. A similar reaction to that of the first procedure occurs, but under the reaction conditions the zwitterionic (dipolar) product eventually loses acetic acid, perhaps by an internal shift of an acetyl group from nitrogen to oxygen.

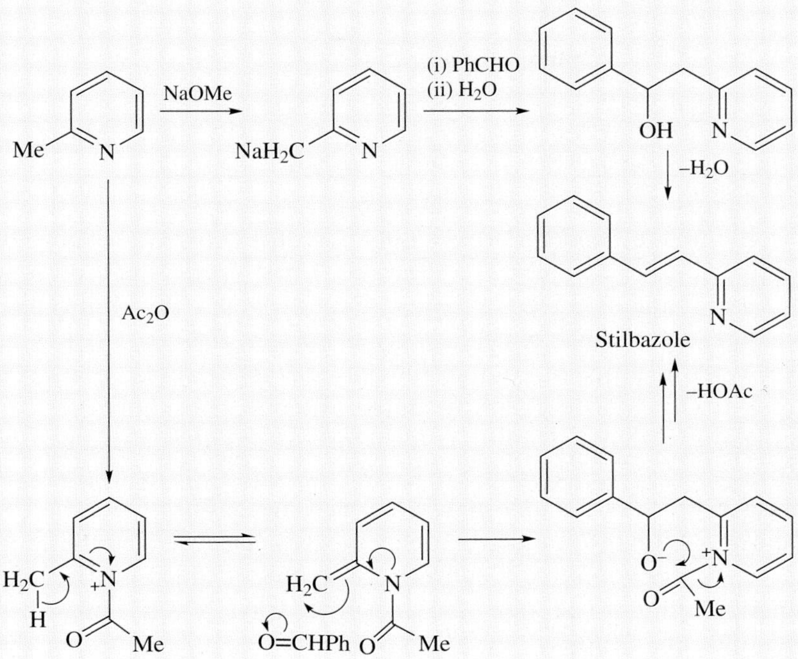

Scheme 2.21

2.7.2 Pyridones (Hydroxypyridines)

Tautomerism

The tautomeric preferences observed for hydroxy and amino heterocycles is a complex subject and not always easily explained.[1,2]

2-Hydroxy- and 4-hydroxypyridines are in tautomeric equilibrium with isomers bearing a carbonyl group (Scheme 2.22). These are called **2-** and **4-pyridones**, respectively. The pyridone forms are favoured in ionic solvents and also in the solid state.

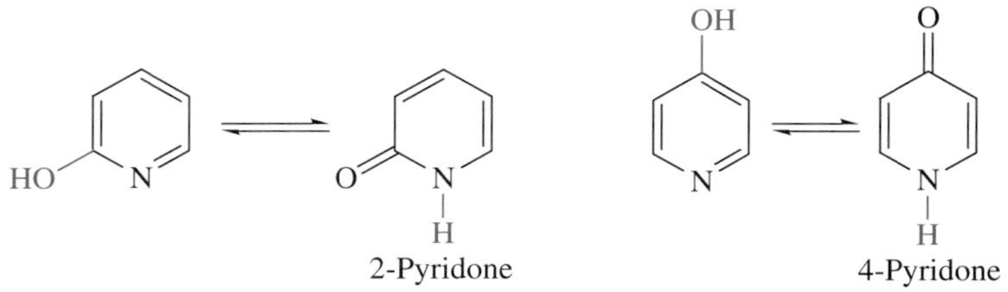

2-Pyridone   4-Pyridone

Scheme 2.22

3-Hydroxypyridine adopts a dipolar (zwitterionic) constitution in polar solvents (Scheme 2.23).

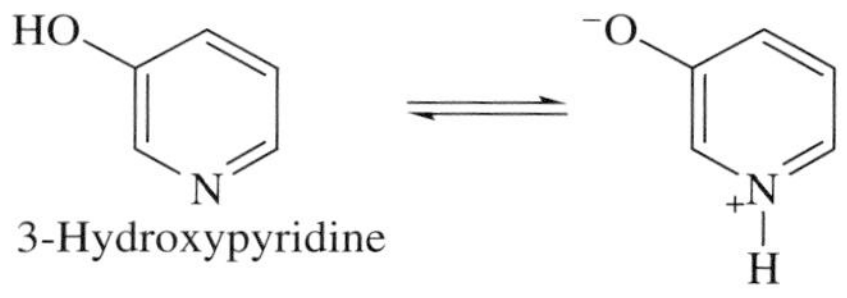

3-Hydroxypyridine

Scheme 2.23

4-Pyridones can be considered to react with electrophiles at C-3 either as enamines or as enols.

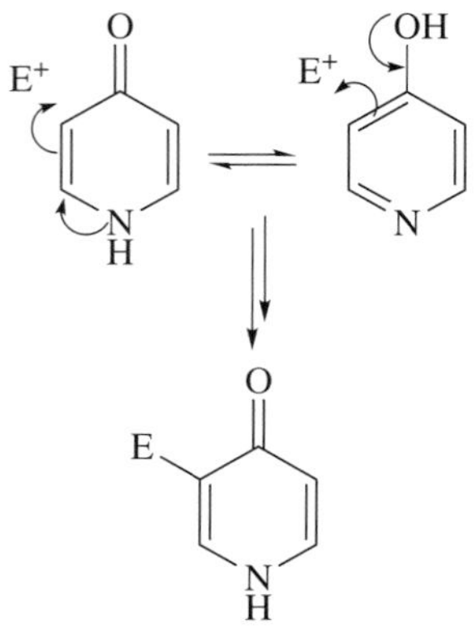

Note that O–Si bonds are stronger than N–C bonds

Reactions

These compounds undergo electrophilic substitution readily, but their chemistry is quite complex; for example, 2-and 4-pyridones, although weak bases (amide-like), protonate at oxygen. Some other electrophiles, particularly those which bond strongly to oxygen, also react at this site. Others react at nitrogen, and acetyl chloride, for example, gives a 1:1 mixture of *O*- and *N*-acetyl-4-pyridones, whereas phosphorus oxychloride, together with phosphorus pentachloride, reacts at oxygen and forms a good leaving group (possibly Cl_2OPO^-), which is then displaced by chloride ion to afford a chloropyridine (Scheme 2.24).

Deprotonation of 2- and 4-pyridones is easily achieved, and the anions react with carbon electrophiles, such as carbon dioxide and trimethylsilylmethyl chloride, at nitrogen, but with trimethylsilyl chloride at oxygen (Scheme 2.25).

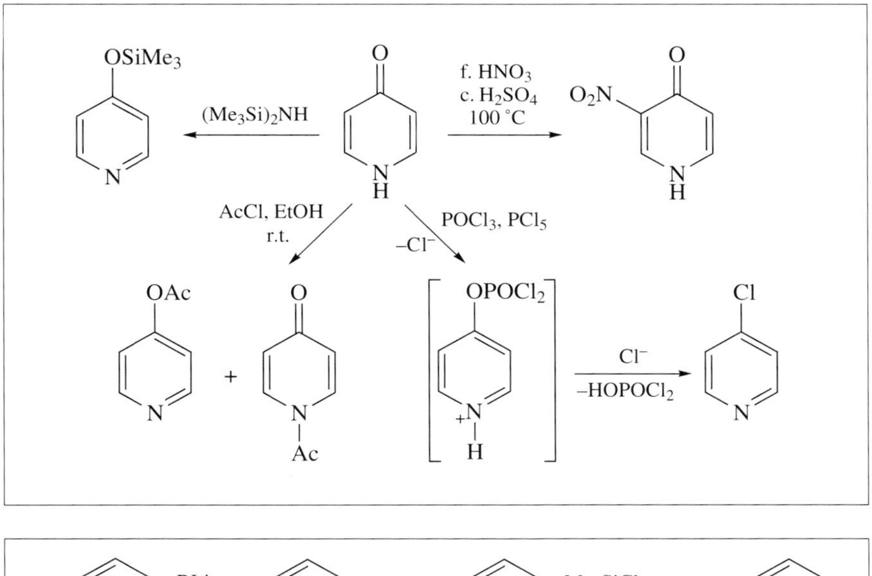

Scheme 2.24

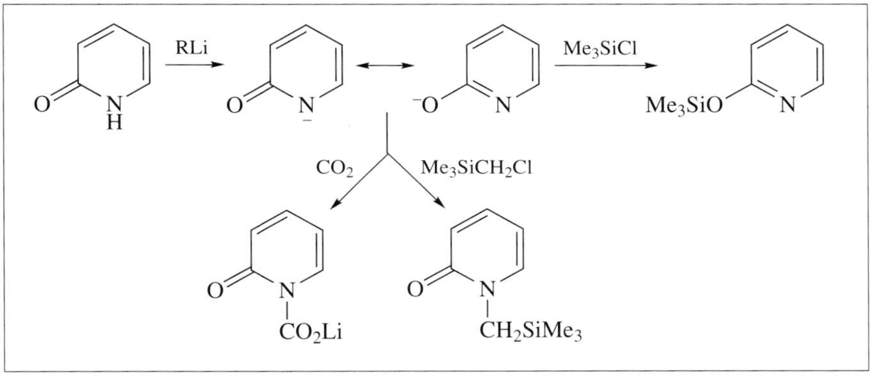

Scheme 2.25

Worked Problem 2.3

Q Devise a synthesis of 2-methoxypyridine from 2-pyridone. Suggest possible mechanisms for the reactions involved

A First, a reaction with phosphorus oxychloride and phosphorus pentachloride can be used to convert 2-pyridone into 2-chloropyridine, and then this compound is subjected to an addition–elimination reaction with methoxide ion (from sodium methoxide) (Scheme 2.26). Note: chloride is a much better leaving 'group' than methoxide.

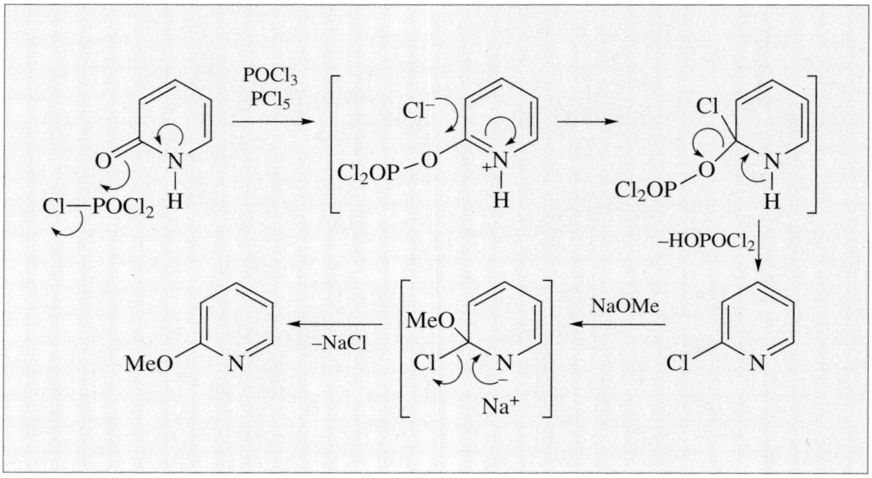

Scheme 2.26

2.7.3 Aminopyridines

All three aminopyridines are known, but although the 2-and 4-aminopyridines are potentially tautomeric with imino forms, they seem to exist as the amino tautomers (Scheme 2.27).

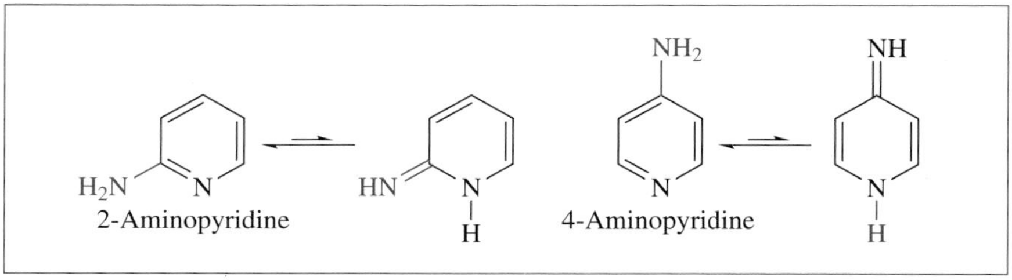

Scheme 2.27

2.7.4 Pyridinium Salts

Nucleophilic Addition

Nucleophilic *addition* readily takes place with pyridinium salts; attack is normally easier at the C-2(6) position, since the inductive effect of the positively charged nitrogen atom is greatest here (Scheme 2.28). When the sites adjacent to the nitrogen are blocked, however, attack occurs at C-4. The products are dihydropyridines.

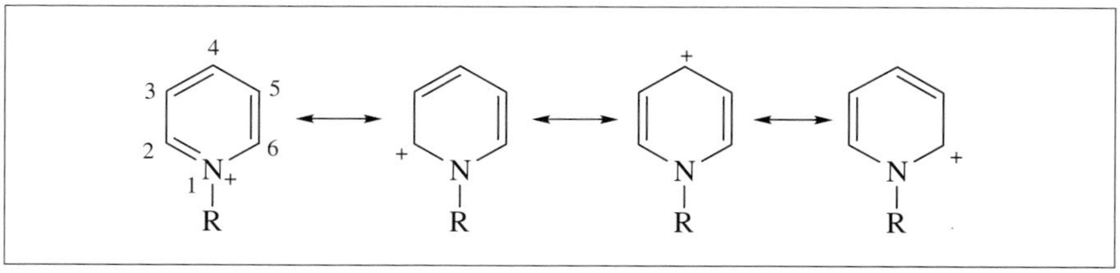

Scheme 2.28

It is not always easy to remove the *N*-substituent after the initial nucle-ophilic addition, although demethylation of *N*-methylpyridinium salts can be achieved by heating in *N*,*N*-dimethylformamide, and *N*-benzyl groups are cleaved by hydrogenolysis (heating over Pd/C in the presence of hydrogen).

Other groups can be bonded to the nitrogen atom and, if these are electron withdrawing, nucleophilic attack at a ring carbon atom is further enhanced. This is particularly useful if the activating group is carbonyl containing, for it can easily be removed at the end of the reac-tion by hydrolysis. Oxidation then leads to a substituted pyridine. However, note that, with electron-withdrawing *N*-substituents, ring opening by certain nucleophiles (*e.g.* cyanide ion) is also facilitated. Such reactions often take place by concerted mechanisms.

Some examples of nucleophilic additions are shown in Scheme 2.29.

As for pyridine, the presence of a potential leaving group already in the ring of the salt at C-2 or C-4 favours preferential attack at this site and facilitates re-aromatiz-ation to a substituted pyridinium salt by loss of the original substituent from the reaction intermediate.

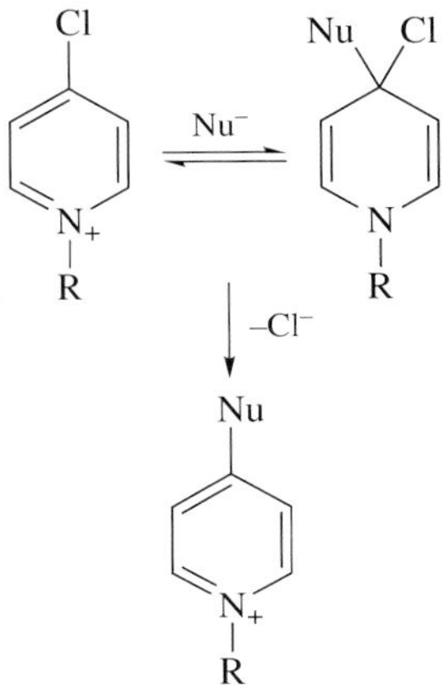

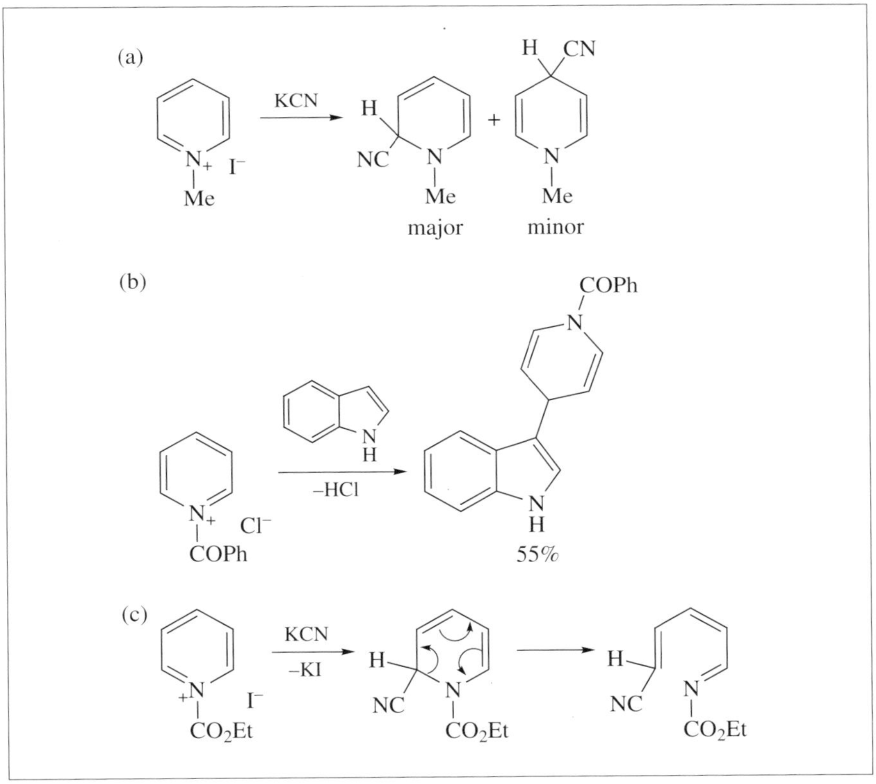

(a)

(b)

In Scheme 2.29b the nucleophile is indole (see Chapter 7), acting as an enamine. It seems likely that the size of the attacking reagent in this case is influential in directing addition solely to C-4. The reaction in Scheme 2.29c exemplifies nucleophilic addition followed by a concerted retro-cyclization process.

Although 2-cyanodihydropyridines are favoured by the addition of potassium cyanide in water to pyridinium salts, an increase in the 4-isomer may occur when a salt like ammonium chloride is added to the reaction mixture.

Scheme 2.29

2.8   Reduced Pyridines

2.8.1 Reduction of Pyridines

Pyridine is difficult to reduce (as is benzene!), but pyridinium salts, *e.g.* alkylpyridinium halides, are *partly* reduced by hydride transfer reagents such as lithium aluminium hydride ($LiAlH_4$) and sodium borohydride ($NaBH_4$). $LiAlH_4$, which must be used in anhydrous conditions, only gives the 1,2-dihydro derivative, but the less vigorous reductant $NaBH_4$ in aqueous ethanol yields the 1,2,5,6-tetrahydro derivative (Scheme 2.30)!

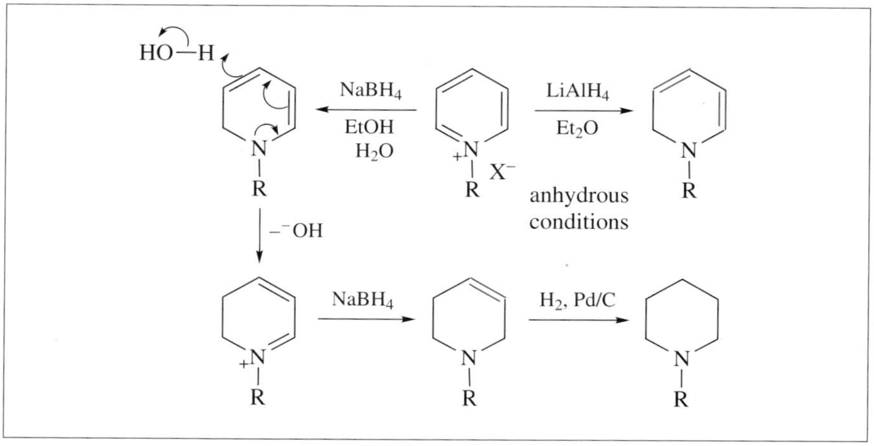

Scheme 2.30

In this case the aqueous conditions permit protonation at C-5 after the first hydride addition, thereby setting up the next hydride transfer. The reduction stops before the ring is fully saturated, however, as the lone pair electrons on the N atom of the 1,2,5,6-tetrahydropyridine are not conjugated with the C=C double bond (*i.e.* further activation by protonation is not possible).

Reduction of 1,2,5,6-tetrahydropyridines to N-alkylpiperidines requires catalytic hydrogenation.

If R = benzyl, this substituent can be hydrogenolysed after, or as a part of, the hydrogenation procedure (*i.e.* without the isolation of the intermediate N-alkylpiperidine).

2.8.2 NAD+ and NADH

NAD^+ is one of Nature's most important oxidizing agents; it can be considered as a biological equivalent of the chromium(VI) ion. NAD^+ is shorthand for **nicotinamide adenine dinucleotide**; it is a co-enzyme, which together with an enzyme is essential for several life-sustaining processes (Box 2.2). On reduction it forms the corresponding 1,4-dihydropyridine, **NADH**. The oxidation of ethanol to acetaldehyde (ethanal) is effected by the enzyme alcohol dehydrogenase and mediated by NAD^+ (Scheme 2.31).

Box 2.2 Nicotinamide Adenine Dinucleotide

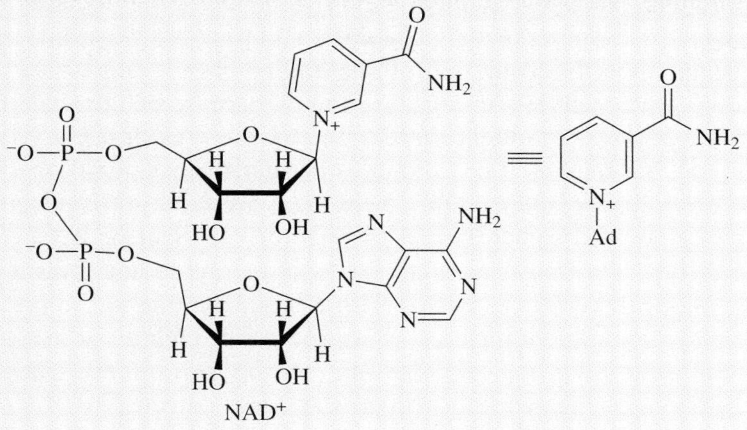

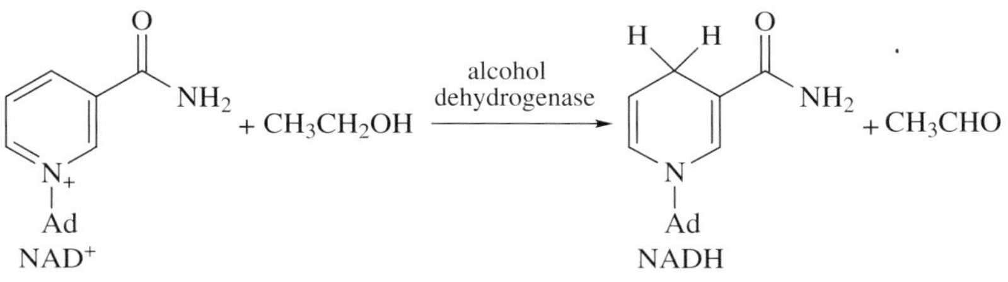

NAD+

Scheme 2.31

Conversely, NADH is one of the important natural reductants, and in yeast cells, for example, it reduces acetaldehyde to ethanol!

2.8.3 Piperidine (Azacyclohexane or Hexahydropyridine)

Synthesis and Reactions

Piperidine is obtained commercially by the catalytic hydrogenation of pyridine over a nickel catalyst at about 200 °C. *N*-Substituted derivatives are formed by reduction of the corresponding pyridinium salts.

Piperidine is a secondary amine (pK_a 11.3; *cf*. diethylamine, pK_a 11.0); it is more basic than pyridine (pK_a 5.2). It is also a good nucleophile, and it is *N*-alkylated by alkyl halides in the presence of potassium carbonate to form first *N*-alkylpiperidines and then quaternary salts.

Hofmann Exhaustive Methylation

The **Hofmann exhaustive N-methylation** procedure, often used in classical structure determinations of alkaloids containing the piperidine nucleus, depends upon the formation of quaternary methyl salts.

Although of much less importance today, because of the availability of NMR spectroscopy and mass spectrometry for structural analysis, the method still has its uses. For example, consider the exhaustive methylation of an 'unknown' piperidine where there are three different groups X, Y and Z located somewhere in the heterocycle (Scheme 2.32). First, treatment of the unknown with an excess of methyl iodide gives a quaternary iodide; then this is reacted with a base. Heating in the presence of the base promotes elimination of a proton, through a concerted E2-type mechanism involving H-3, causing the ring to fragment. Then, the whole process is repeated until trimethylamine is detected.

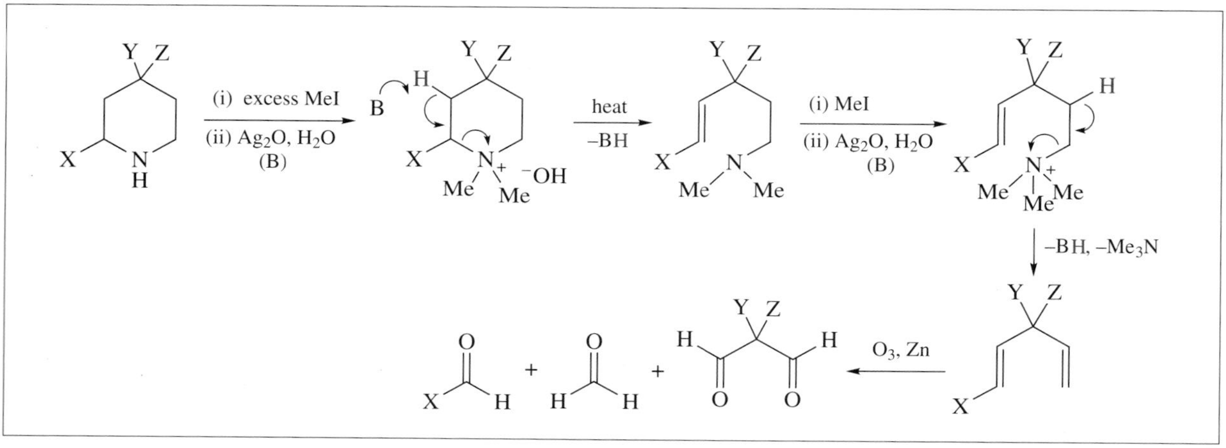

Originally, moist silver oxide (expensive!) was used as the base, although an alkali or an alkali-bound ion exchange resin are alternatives.

In practice, the formation of trimethylamine is obvious from its smell, or by red litmus paper turning blue when applied to the top of the condenser.

Scheme 2.32

For our 'unknown' piperidine the other product is a penta-1,4-diene, the constitution of which can now identified by ozonolysis (in the presence of a reducing agent such as Zn dust, or triphenylphosphine, to prevent the oxidation of the ozonolysis fragments). Here ozonolysis will yield formaldehyde (methanal) (an indication that C-6 is unsubstituted) plus another aldehyde, XCHO (showing that C-2 bears the group X). The other product is a propanedial in which the groups Y and Z must occupy the central carbon. Clearly then, C-4 is substituted by both Y and Z.

The above example was selected because the penta-1,4-diene cannot isomerize to a more conjugated isomer; however, in many other examples this is not the case and, for example, a Hofmann exhaustive methylation reaction upon piperidine itself eventually leads to penta-2,4-diene (Scheme 2.33).

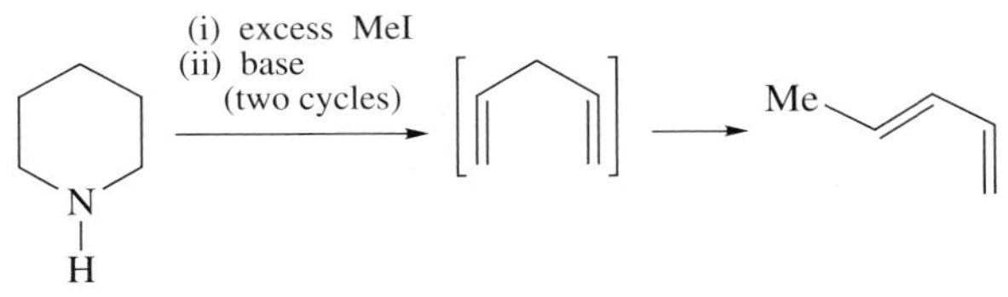

Scheme 2.33

Furthermore, when a group bearing an α-hydrogen atom is present at C-2, proton loss from this substituent may occur, rather than loss of the C-3 ring proton. This gives rise to a double bond exocyclic to the original heterocycle (Scheme 2.34).

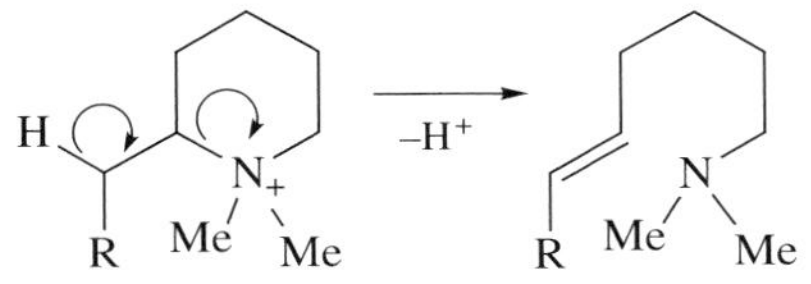

Scheme 2.34

Summary of Key Points

1. Pyridine is both aromatic and basic, and combines with electrophiles at nitrogen, disfavouring further attack at the carbon ring atoms.

2. When electrophilic substitution does occur, it does so at C-3. However, such reactions are either very slow, or may proceed through complex mechanisms.

3. In order to overcome problems with reactivity at C-2 and C-4, pyridine N-oxide can be employed as the substrate for electrophilic substitution.

4. Nucleophilic attack on pyridine is favoured at C-2(6) and C-4; the selectivity of such reactions is influenced by the presence of a good leaving group at the site of attack.

5. Pyridinium salts are susceptible to addition reactions with nucleophiles and to reduction.

6. Reduction can lead to dihydro-, tetrahydro- and hexahydropyridines (piperidines), depending upon the nature of the reagents and the reaction conditions.

Problems

1. Explain why 2-chloropyridine reacts more readily with nucleophiles than pyridine.

2. Indicate why pyridine *N*-oxide can react with both nucleophiles and electrophiles.

3. When 3-chloropyridine is reacted with sodamide in liquid ammonia, both 3- and 4-aminopyridines are formed. What common intermediate is involved in these reactions?

4. Provide syntheses of (a) 2-ethylpyridine from 2-methylpyridine and (b) 1-benzyl-*trans*-3,4-dibromopiperidine from 1-benzylpyridinium bromide.

5. Suggest a synthesis of compound **2.2**.

2.2

Note: the substituents at C-3 and C-5 really are different!
Hint: a 'simple' Hantzsch procedure cannot be used since mixed products would form (see Section 2.6.1, p. 28).

6. A compound could be either 4-methoxycarbonyl-2,5-diphenylpiperidine or 4-methoxycarbonyl-3,5-diphenylpiperidine. How could you determine which formulation is correct, using the Hofmann exhaustive methylation procedure followed by 'reductive' ozonolysis?

References

1. J. Elguero, C. Marzin, A. R. Katritzky and P. Linda, *The Tautomerism of Heterocycles*, in *Adv. Heterocycl. Chem., Suppl. 1*, 1976.
2. A. R. Katritzky, M. Karelson and P. A. Harris, *Prototropic Tautomerism of Heterocyclic Compounds*, in *Heterocycles*, 1991, **32**, 329.

Further Reading

E. Klinsberg (ed.), *Pyridine and its Derivatives,* in *Chemistry of Heterocyclic Compounds*, suppl. 1–4, ed. A. Weissberger and E. C. Taylor, Wiley-Interscience, New York, 1960–1964.

V. Eisner and J. Kutham, *The Chemistry of Dihydropyridines*, in *Chem. Rev.*, 1972, **72**, 1.

C .K. McGill and A. Rappa, *Advances in the Chichibabin Reaction*, in *Adv. Heterocycl. Chem.*, 1988, **44**, 2.

A. R. Katritzky and J. M. Lagowski, *Heterocyclic N-Oxides*, Methuen Press, London, 1967.

H. Vorbrüggen, *Advances in the Amination of Nitrogen Heterocycles*, in *Adv. Heterocycl. Chem.*, 1990, **49**, 117.

A. Albini and S. Pietra, *Heterocyclic N-Oxides*, CRC Press, Boca Raton, Florida, 1991.

3
Benzopyridines

Aims

By the end of this chapter you should understand:

- How the presence of the benzene nucleus influences the chemistry of quinolines and isoquinolines compared to that of pyridine
- The main methods used for the synthesis of quinolines, isoquinolines and their derivatives

3.1 Introduction

The naming of benzo-fused heterocycle is discussed in Section 1.2.

There are two neutral benzopyridines, **quinoline** (benzo[*b*]pyridine) and **isoquinoline** (benzo[*c*]pyridine), together with the **quinolizinium cation** (benzo[*a*]pyridinium) (Box 3.1), formed by a ring fusion that utilizes the nitrogen atom.

Box 3.1 Benzopyridines

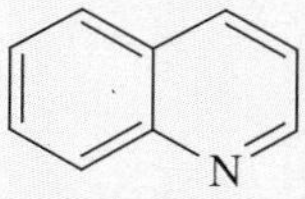

Quinoline
(benzo[*b*]pyridine)

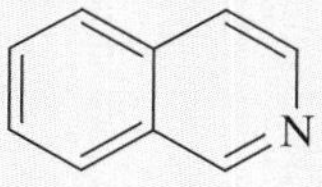

Isoquinoline
(benzo[*c*]pyridine)

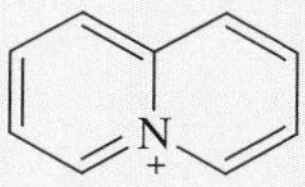

Quinolizinium cation
(benzo[*a*]pyridinium)

Quinolines and isoquinolines are very important because their derivatives, a large proportion of which are **alkaloids**, show useful biological effects. Indeed, the medicinal properties of the plants that biosynthesize these alkaloids have been recognized for centuries, long before the nature of the compounds responsible was known. Two such natural products are the anti-malarial agent **quinine**, from the bark of Cinchona trees, and **morphine**, a tetrahydroisoquinoline derivative, found in the latex from poppy seed capsules. Morphine and its di-*O*-acetyl derivative, **heroin**, are still used to control severe pain, despite being addictive drugs.

3.2 Quinoline

3.2.1 Molecular Structure and General Properties

Quinoline is a base since, as for pyridine, the lone pair of electrons on the nitrogen atom is not utilized in its internal resonance. Although it is an aromatic compound, the valence bond description of quinoline shows two of the neutral contributors, A and C (see Scheme 3.1), to the resonance hybrid as quinonoid in character, whereas in B either the carbocycle or the heterocycle must exist in the form of a 1,3-diene. The presence of the pyridine nucleus is reflected by the inclusion of doubly charged canonical forms.

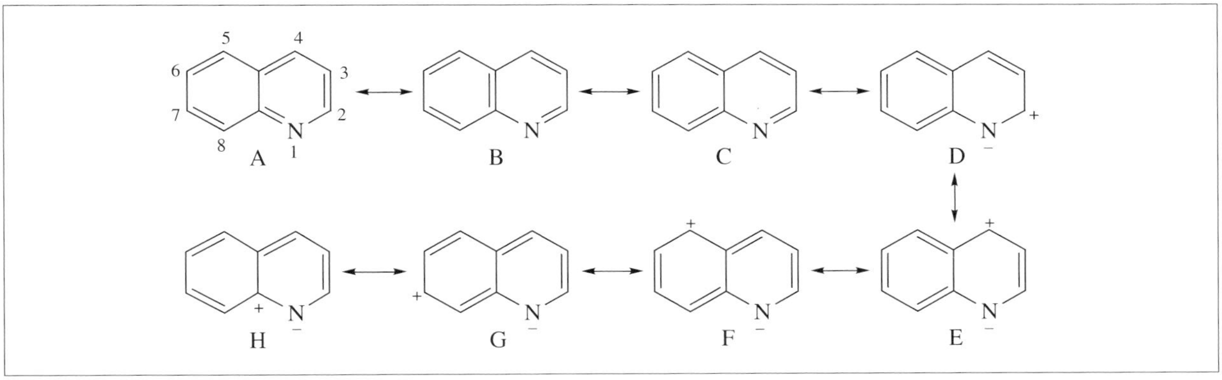

However, the representations F to H involve disruption of both monocyclic π systems simultaneously. It follows that these are of higher energy, and they contribute very much less to the overall description of the molecule than do the alternatives D and E that affect only the pyridine system.

Scheme 3.1

The bond lengths of quinoline, which are irregular, support the resonance description; thus, the 1,2-, 5,6- and 7,8-linkages are shorter than that of the carbon–carbon bond in benzene (more double bond character!). There is also a dipole of 2.19 D directed towards the nitrogen atom.

3.2.2 Electrophilic Substitution

Acids and Lewis acids react with quinoline at the basic nitrogen atom to form quinolinium salts, and there is a question over the nature of the substrate for **electrophilic attack**, *i.e.* is it quinoline or the quinolinium cation? The answer is not a simple one and appears to depend upon the reagents and reaction conditions. Thus, whereas acetyl nitrate at 20 °C gives mainly 3-nitroquinoline (Scheme 3.2), fuming nitric acid in concentrated sulfuric acid containing sulfur trioxide at 15–20 °C yields a mixture of 5-nitroquinoline (35%) and 8-nitroquinoline (43%) (Scheme 3.3). In the case of acetyl nitrate, the reaction may proceed by the 1,4-addition of the reagent to quinoline, followed by electrophilic attack upon the 1,4-dihydro derivative.

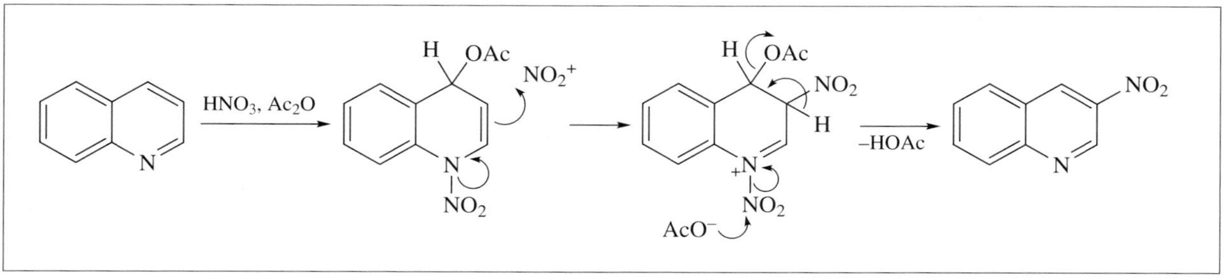

Scheme 3.2

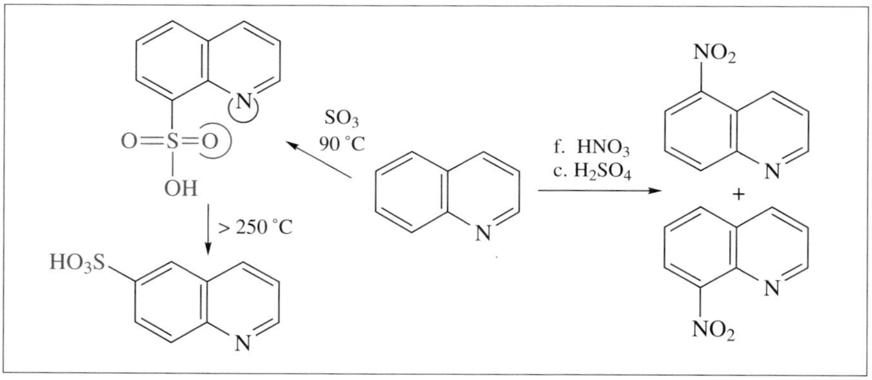

Scheme 3.3

Although not shown, the nitrogen atom of the sulfonic acid is probably protonated, or even sulfonated, under the reaction conditions, exaggerating the steric problems experienced by the substituent at C-8. The mechanism of the desulfonation of aromatic sulfonic acids occurs *via* the reverse of the sulfonation process.

However, the rate of nitration of quinoline in 80–99% sulfuric acid is of the same order as that of *N*-methylquinolinium salts, suggesting that here the quinolinium cation may be the target for attack.

Sulfonation with oleum at 90 °C affords mainly the 8-sulfonic acid, but as this product is sterically hindered, at higher temperatures it rearranges into the 6-sulfonic acid (Scheme 3.3). This rearrangement is similar to that shown by naphthalene-1-sulfonic acid, the kinetic sulfonation product of naphthalene, which isomerizes on heating into the thermodynamically favoured (less hindered) 2-isomer.

Alkyl and acyl halides react directly with quinoline to give *N*-alkyl- or *N*-acylquinolinium salts (Scheme 3.4). Whereas the *N*-alkyl salts are stable and can often be isolated as crystalline solids, the *N*-acyl analogues are unstable and undergo rapid hydrolysis in moist air or in aqueous solution.

The origin of the instability of *N*-acylpyridinium salts and their quinolinium counterparts is the proximity of the positive charge on nitrogen to the positive end of the dipole of the carbonyl group.

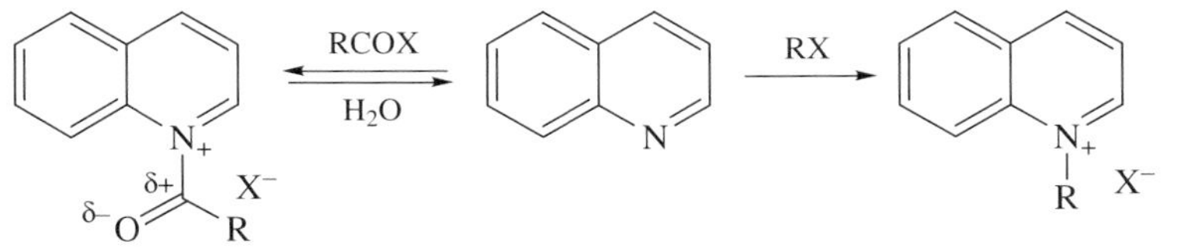

Scheme 3.4

3.2.3 Nucleophilic Addition/Substitution

N-Acyl- or *N*-sulfonylquinolinium salts can be trapped by cyanide ion to form what are known generally as **Reissert adducts**. The easy removal of the *N*-substituent in a subsequent reaction with a base provides access to 2-cyanoquinoline (Scheme 3.5).

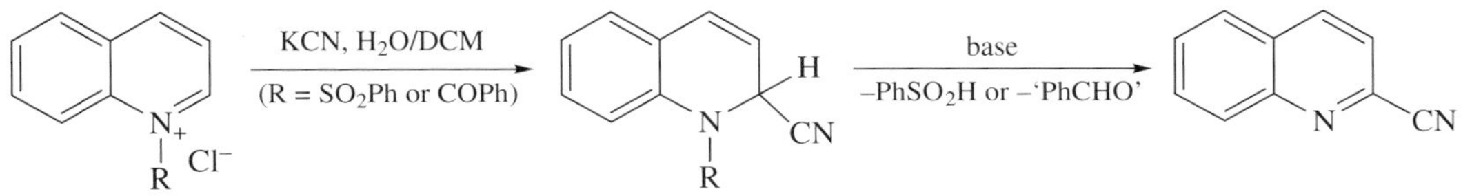

Scheme 3.5

There is a strong similarity between the reactions of pyridines and quinolines towards nucleophiles. Addition occurs mainly at C-2, giving 1,2-dihydroquinolines, but the locus of the reaction can be diverted to C-4, particularly if there is a good leaving group located at this position.

In a **Chichibabin-type reaction** (see Section 2.4.2), quinoline reacts with potassamide (KNH_2) in liquid ammonia at –70 °C to give 2-amino-1,2-dihydroquinoline and this is oxidized by potassium permanganate [manganate(VII)] at the same temperature to yield 2-aminoquinoline (Scheme 3.6). If the temperature is allowed to increase to –45 °C the adduct rearranges into 4-amino-1,4-dihydroquinoline, and upon oxidation this product gives 4-aminoquinoline.

3.2.4 Synthesis

Skraup Synthesis

The commonest approach is the **Skraup synthesis**. Here a mixture of glycerol (propane-1,2,3-triol), aniline (phenylamine), sulfuric acid, nitrobenzene and ferrous [iron(II)] sulfate are heated together (Scheme

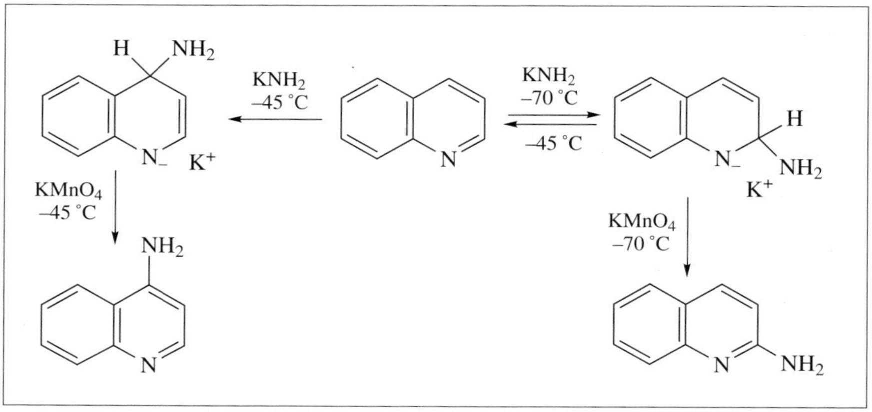

Scheme 3.6

3.7). The last reagent is added as a moderator to prevent a 'runaway reaction'; nitrobenzene, or an alternative oxidant (iodine or chloroanil are often recommended), is required to convert the product, 1,2-dihydroquinoline, into quinoline.

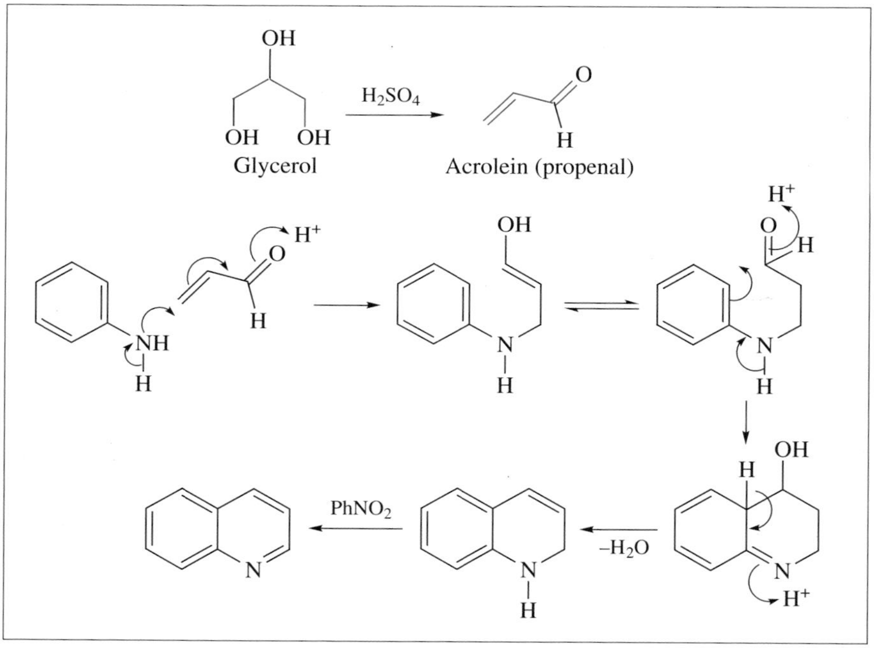

Scheme 3.7

The Skraup procedure is applicable to quinoline derivatives substituted in the benzene ring, providing these substituents are not strongly electron withdrawing.

Friedländer Synthesis

Another approach, the Friedländer synthesis, is to condense a 2-aminophenyl ketone with an aldehyde or ketone that contains a methylene unit (to permit enolization and subsequent aromatization), in contact with either an acid or a base as the catalyst (Scheme 3.8).

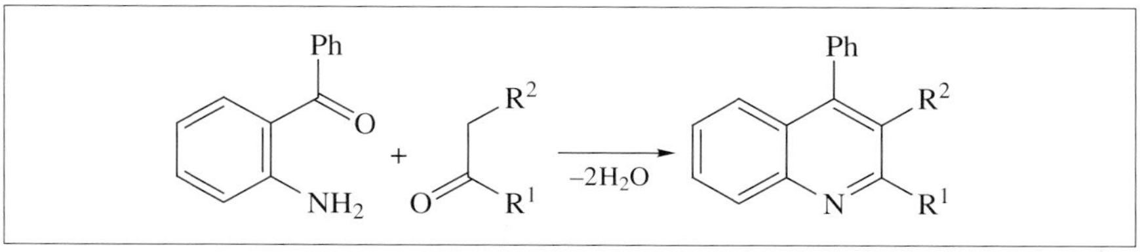

Scheme 3.8

3.2.5 Important Derivatives

Methylquinolines

Like the corresponding methylpyridines, 2- and 4-methylquinolines can be deprotonated by a base, such as sodium methoxide, forming resonance-stabilized anions (Scheme 3.9). The latter are useful in synthesis, providing nucleophilic reagents that allow extension of quinoline side chains through reactions with appropriate electrophiles. Activation of the 2-methyl group can also be achieved by the use of acetic anhydride (the same type of process occurs with 2-methylpyridine, Section 2.7.1, Worked Problem 2.3).

Scheme 3.9

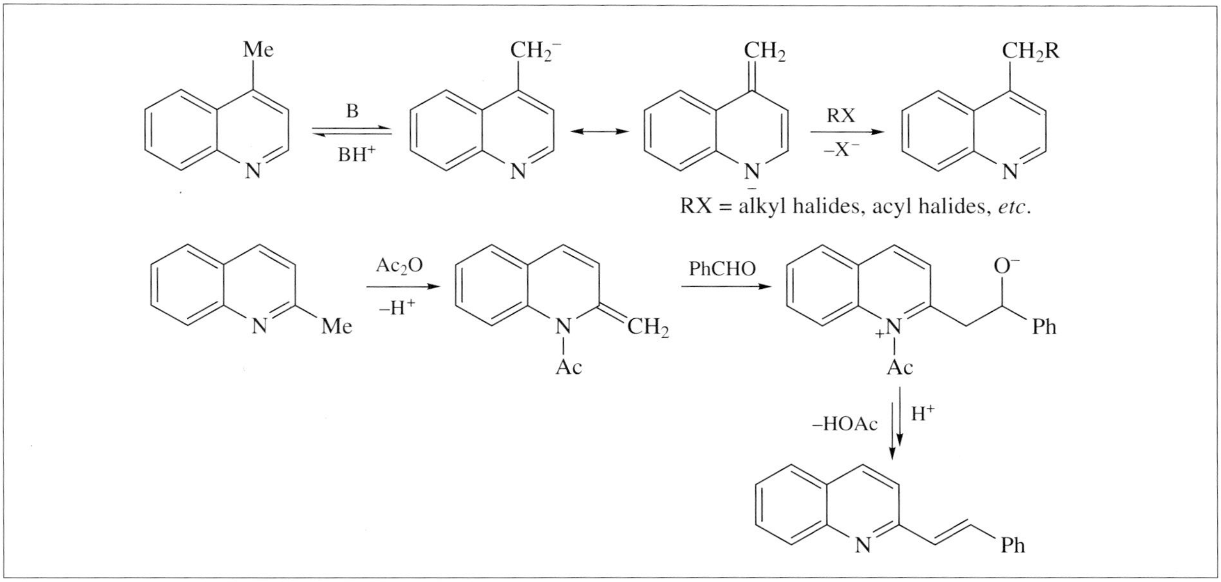

RX = alkyl halides, acyl halides, *etc.*

Aminoquinolines

These compounds exist as the amino tautomers and normally the imino forms are not observed. All react with acids at the heterocyclic nitrogen atom, giving salts. The protonated forms of both 2- and 4-aminoquinolines are resonance hybrids (Scheme 3.10), but 4-aminoquinoline is more basic than 2-aminoquinoline, possibly because the nitrogen atoms that carry the charge between them in the corresponding cation are more widely separated.

Suitable imines are commonly in equilibrium with enamine tautomers:

imime

enamine

Scheme 3.10

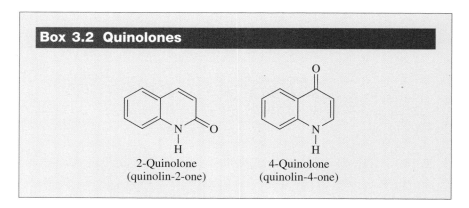

2-Aminoquinoline   2-Aminoquinolinium cation

4-Aminoquinoline   4-Aminoquinolinium cation

Quinolones (Quinolinones)

Although hydroxyquinolines in which the hydroxy substituent is present in the benzene ring are known and show phenolic activity, 2- and 4-hydroxyquinoline exist solely as the quinolone tautomers (Box 3.2).

Box 3.2 Quinolones

2-Quinolone
(quinolin-2-one)

4-Quinolone
(quinolin-4-one)

Conrad–Limpach–Knorr Synthesis

Quinolones are obtained in the Conrad–Limpach–Knorr synthesis, which is subject to either kinetic or thermodynamic control, when aniline is reacted with a 3-keto ester (Scheme 3.11a). At room temperature the more reactive keto group combines with the aniline nitrogen atom, leading to an enamino ester: the kinetic product. Cyclization of this product to a 4-quinolone requires heating at 250 °C.

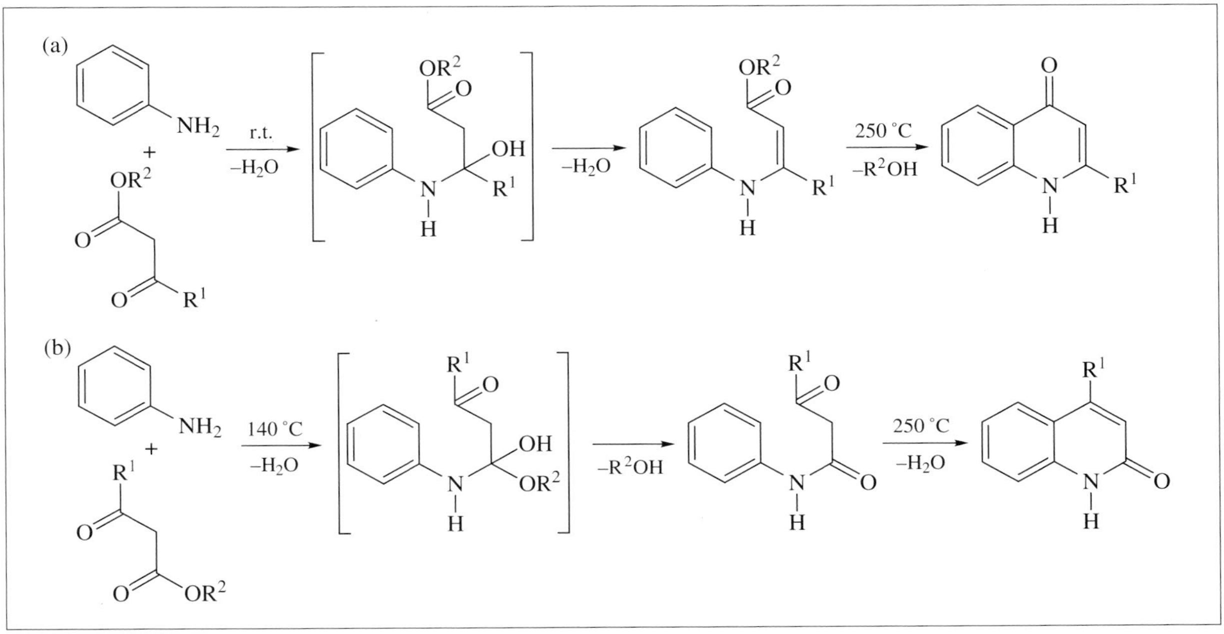

Alternatively, when the reaction is carried out at 140 °C, the thermodynamically preferred amido ketone is formed (Scheme 3.11b), even though the less reactive ester group of the keto ester is the locus of the initial nucleophilic attack. Ring closure of the amino ketone then affords the 4-quinolone.

Scheme 3.11

Worked Problem 3.1

Q Devise a synthesis of 1-benzyl-1,2,3,4-tetrahydroquinoline from quinoline.

A By reacting quinoline with benzyl bromide in an inert solvent, such as diethyl ether, the corresponding quinolinium salt is obtained (Scheme 3.12). Treatment of this product with sodium borohydride leads to 1-benzyl-1,2-dihydroquinoline. Since the double bond is not conjugated with the nitrogen atom, except through the benzene ring, it is now necessary to reduce this bond by catalytic hydrogenation under mild conditions (selected so that hydrogenolysis of the benzyl group does not occur). Note: it would require much higher pressure to reduce the benzene ring.

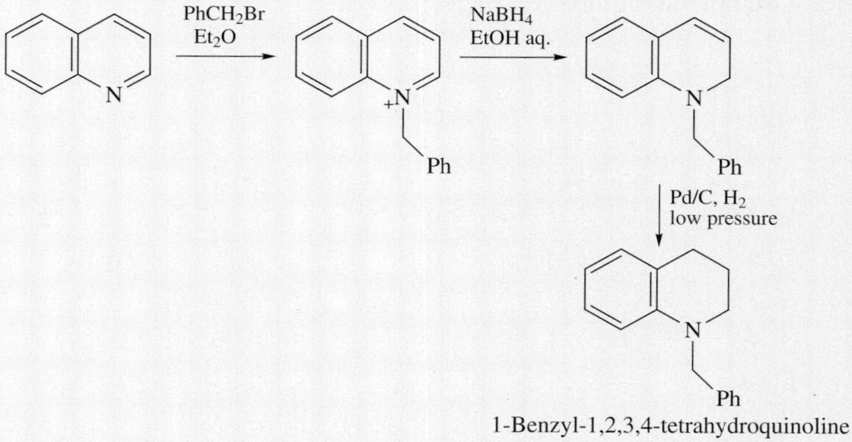

1-Benzyl-1,2,3,4-tetrahydroquinoline

Scheme 3.12

3.3   Isoquinoline

3.3.1 Resonance Structure

The valence bond description of isoquinoline is similar to that of quinoline (Scheme 3.13). Only A, B, C and D are of major importance in describing the molecule, as the contributors E, F, G and H involve a disruption of the π-systems of both rings.

Scheme 3.13

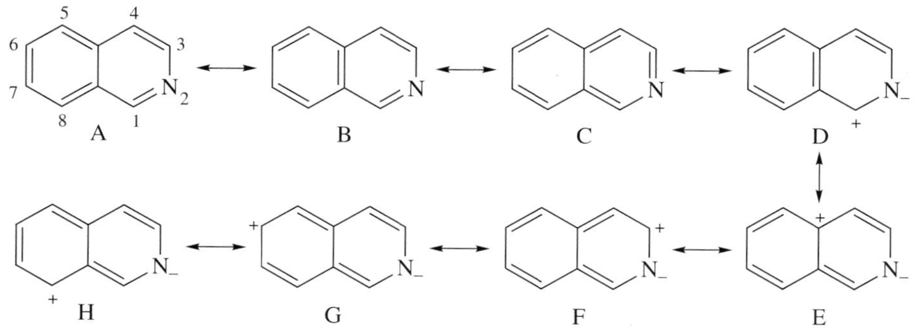

3.3.2 Reactions with Electrophiles

Isoquinoline, like quinoline, is protonated and alkylated at the nitrogen atom, but electrophilic substitution in the benzene ring is also easily achieved (Scheme 3.14). Sulfonation with oleum gives mainly the 5-sulfonic acid, but fuming nitric acid and concentrated sulfuric acid at 0 °C produce a 1:1 mixture of 5- and 8-nitroisoquinolines. Bromination in the presence of aluminium trichloride at 75 °C gives a 78% yield of 5-bromoisoquinoline.

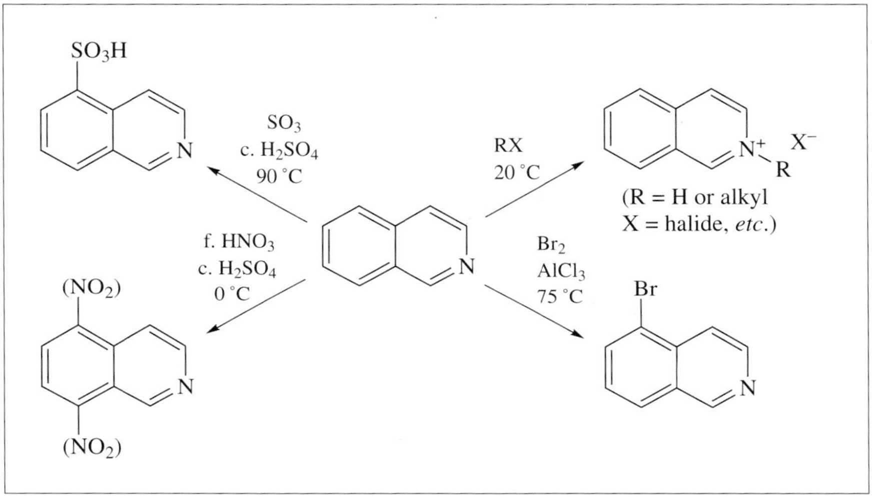

Scheme 3.14

3.3.3 Reduction and Reactions with Nucleophiles

Nucleophilic addition takes place at C-1, and this is considerably enhanced if the reaction is carried out upon an isoquinolinium salt. Reduction with lithium aluminium hydride [tetrahydroaluminate(III)] in THF (tetrahydrofuran), for example, gives a 1,2-dihydroisoquinoline (Scheme 3.15). These products behave as cyclic enamines and if isoquinolinium salts are reacted with sodium borohydride [tetrahydroboronate(III)] in aqueous ethanol, further reduction to 1,2,3,4-tetrahydroisoquinolines is effected through protonation at C-4 and then hydride transfer from the reagent to C-3.

The cyanide anion adds to C-1 in 2-benzoylisoquinolinium salts in water/DCM (dichloromethane), forming Reissert compounds; then, just like their quinoline counterparts (see Section 3.2.3), the adducts can be deprotonated by a base with the loss of the *N*-substituent and the formation of a 1-cyanoisoquinoline (Scheme 3.16).

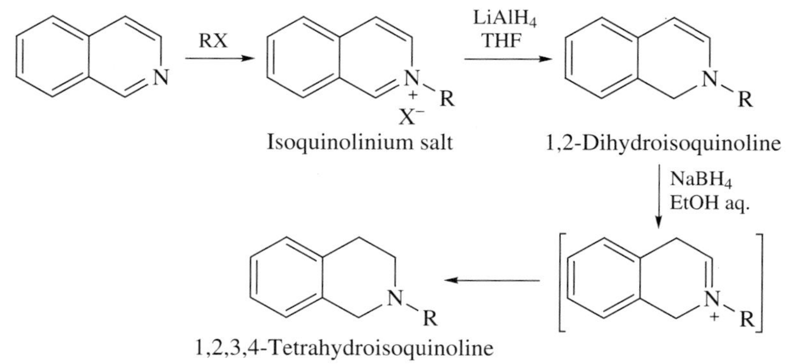

Scheme 3.15

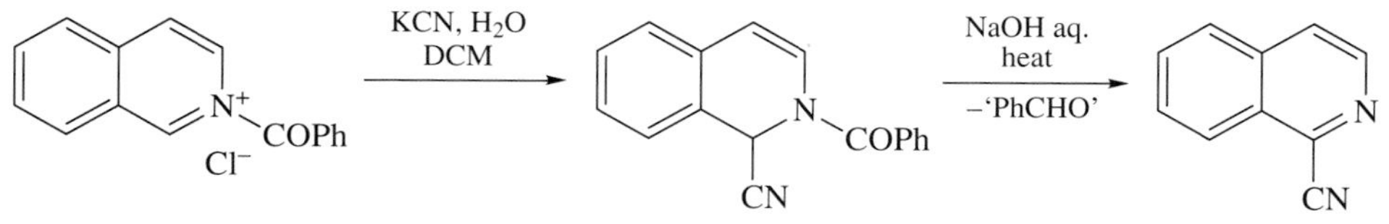

Scheme 3.16

3.3.4 Synthesis

The biological properties of many derivatives have ensured the development of a number of syntheses providing access to all types of isoquinolines, both natural and man made. Three important routes are the **Bischler–Napieralski**, **Pictet–Spengler** and **Pomeranz–Fritsch** procedures.

Bischler–Napieralski Synthesis

This method is very useful for the construction of 1-substituted 3,4-dihydroisoquinolines, which if necessary can be oxidized to isoquinolines. A β-phenylethylamine (1-amino-2-phenylethane) is the starting material, and this is usually preformed by reacting an aromatic aldehyde with nitromethane in the presence of sodium methoxide, and allowing the adduct to eliminate methanol and give a β-nitrostyrene (1-nitro-2-phenylethene) (Scheme 3.17). This product is then reduced to the β-phenylethylamine, commonly by the action of lithium aluminium hydride. Once prepared, the β-phenylethylamine is reacted with an acyl chloride and a base to give the corresponding amide (R^1 = H) and then this is cyclized to a 3,4-dihydroisoquinoline by treatment with either phosphorus pentoxide or phosphorus oxychloride (Scheme 3.18). Finally, aromatization is accomplished by heating the 3,4-dihydroisoquinoline over palladium on charcoal.

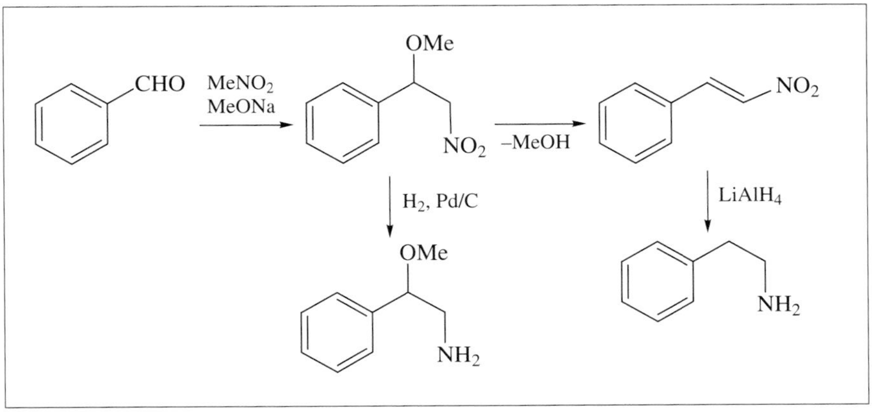

Scheme 3.17

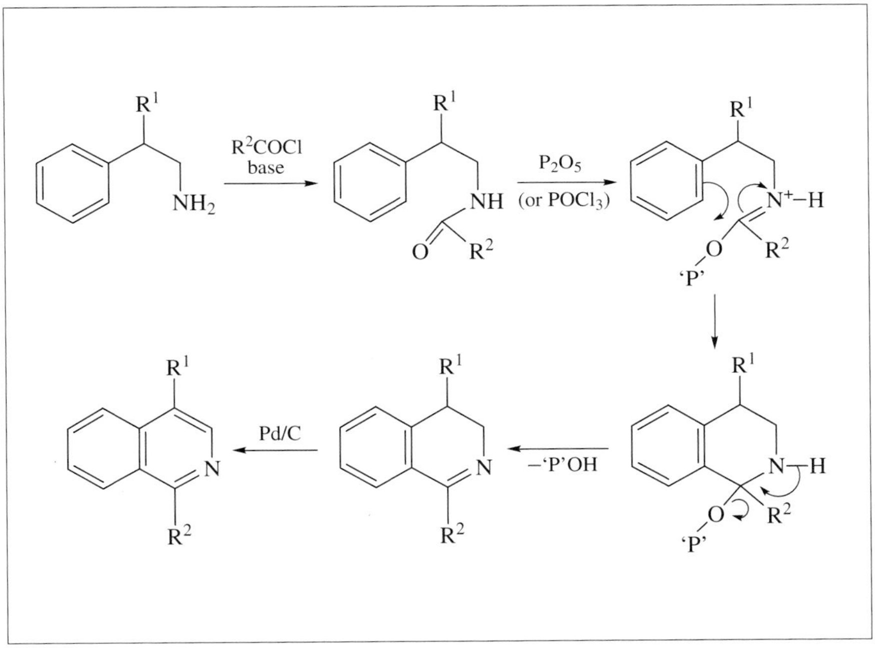

Scheme 3.18

Alternatively, a β-methoxy-β-phenylethylamine can be used to cir-
cumvent the oxidation step after the conventional Bischler–Naperialski
cyclization. Here, when treated with the phosphorus reagent the amide
(R^1 = OMe) undergoes both cyclization and the elimination of methanol
to give the isoquinoline (R = H) directly. This is known as the
Pictet–Gams modification of the Bischler–Napieralski synthesis.

There are numerous other
reagents that can be used in the
cyclization step of the
Bischler–Naperialski synthesis,
but P_2O_5 or $POCl_3$ are probably
the most common.

Pictet–Spengler Synthesis

The mechanistically similar Pictet–Spengler synthesis is also much used
for the preparation of 1,2,3,4-tetrahydroisoquinolines, starting from a

β-phenylethylamine and an aldehyde (Scheme 3.19). The reaction intermediate is an imine which, provided the benzene ring contains electron-donating groups, often ring closes under very mild conditions. Indeed, cyclization can occur under physiological conditions, and in Nature this is an important step in the biosynthesis of many tetrahydroisoquinoline alkaloids.

Scheme 3.19

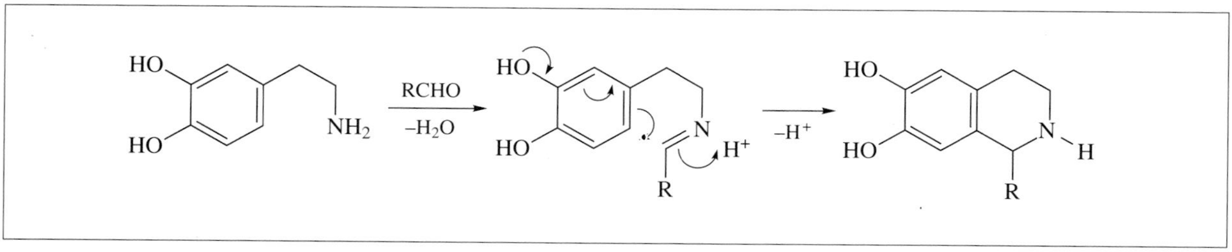

Pomeranz–Fritsch Synthesis

Acetals are formed by reacting an aldehyde or a ketone with an alcohol (2 equivalents) in the presence of a catalytic amount of acid. Water is released and this is trapped out to facilitate the forward reaction. Acetals are stable to base but decompose into their starting materials when treated with aqueous acid:

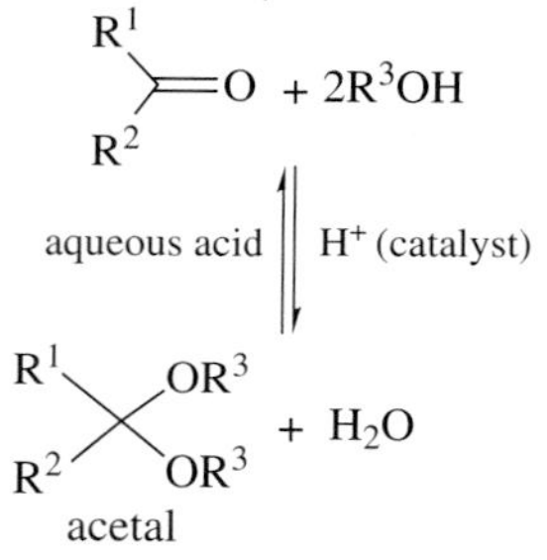

Scheme 3.20

Whereas both the previous two routes depend upon a cyclization of the benzene ring to what becomes C-1 of the heterocycle, the key step in the Pomeranz–Fritch synthesis is the formation of a bond to C-4. A benzaldehyde is the starting material, and it is reacted with an amino-acetaldehyde dialkyl acetal to form an imine, which is then cyclized directly under relatively severe acidic conditions (*e.g.* conc. H_2SO_4 at 100 °C) to give the isoquinoline (Scheme 3.20). Although the Pomeranz–Fritch ring-closure conditions permit the cyclization of unsubstituted imines, the reaction is accelerated greatly if electron-donating groups are present in the benzene ring.

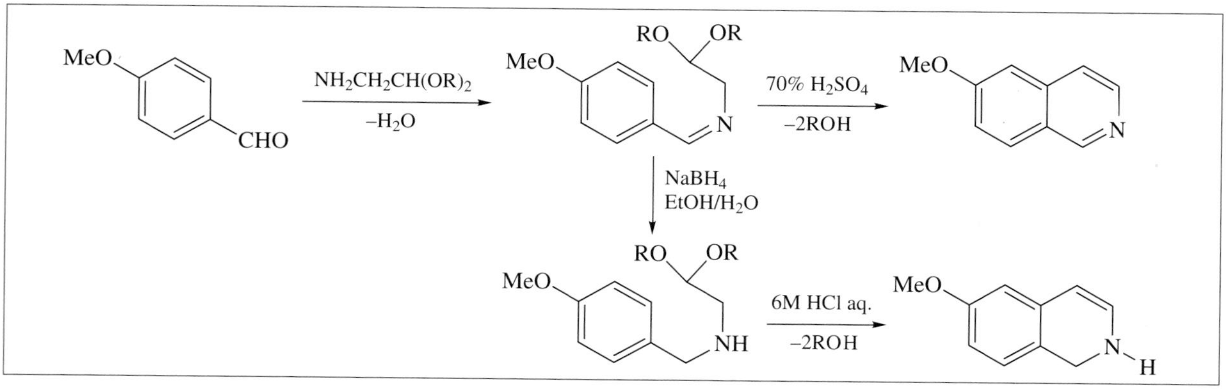

Through a slight modification the Pomeranz–Fritsch synthesis can be made particularly useful for the preparation of 1,2-dihydroisoquinolines. The imine is first reduced with sodium borohydride in 98% ethanol to the corresponding benzylamine, prior to cyclization, by treatment with 6 M hydrochloric acid. When electron-donating groups (such as a methoxyl) are present in the aromatic unit of the benzylamine, the ring-closure step occurs at room temperature to give a 1,2-dihydroisoquino-line. As 1,2-dihydroisoquinolines are unstable in air it is customary to carry out the reaction under an atmosphere of oxygen-free nitrogen.

An advantage of the modified Pomeranz–Fritsch synthesis is that the 1,2-dihydroisoquinolines can be reacted *in situ* with electrophiles, yielding 1,4-dihydroisoquinolinium salts that react with nucleophiles at C-3 (see Section 3.3.3). Such a 'single pot' procedure can be used to form complex 1,2,3,4-tetrahydroisoquinolines.

For convenience the imine is drawn here with a *Z* configuration, but in reality the favoured stereochemistry is *E*; thus, inversion of double bond geometry is necessary prior to ring closure. This requirement contributes to the relative severity of the conditions needed for cyclization compared to those needed for the ring closure of the corresponding benzylamines, where there is free rotation about the C–N bond.

Worked Problem 3.2

Q Suggest a synthesis of the tetracycle **3.1** from 4-methoxyben-zaldehyde and 1-(2-bromoethyl)-3-methoxybenzene.

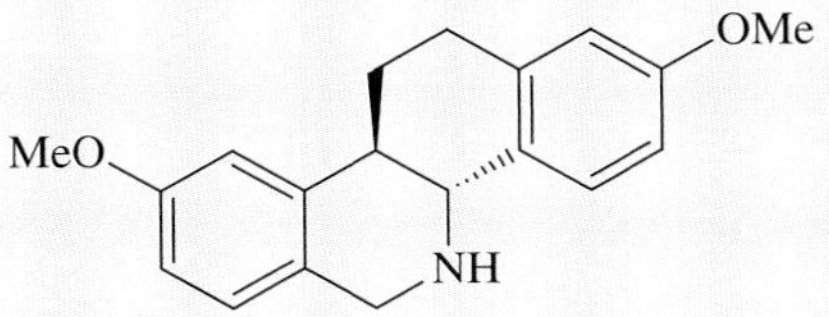

3.1

A Once the 1,2-dihydroisoquinoline is formed by a Pomeranz–Fritsch synthesis between the reduced imine, from 4-methoxybenzaldehyde and aminoacetaldehyde diethyl acetal, it is combined directly with the 1-(2-bromoethyl)-3-methoxybenzene in a tandem 'two-steps-in-one' procedure (Scheme 3.21). First the compound acts as an enamine and combines with the alkyl bromide at C-4, and then the methoxylated phenyl ring of the intermediate reacts with the iminium unit at C-3 to form the tetracycle.

Here is another case where nucleophilic activation by a methoxyl group is necessary for cyclization, and this initiates the attack upon the heterocycle at C-3 and hence the formation of the *trans*-fused tetracycle (this is more stable than the *cis* alternative).

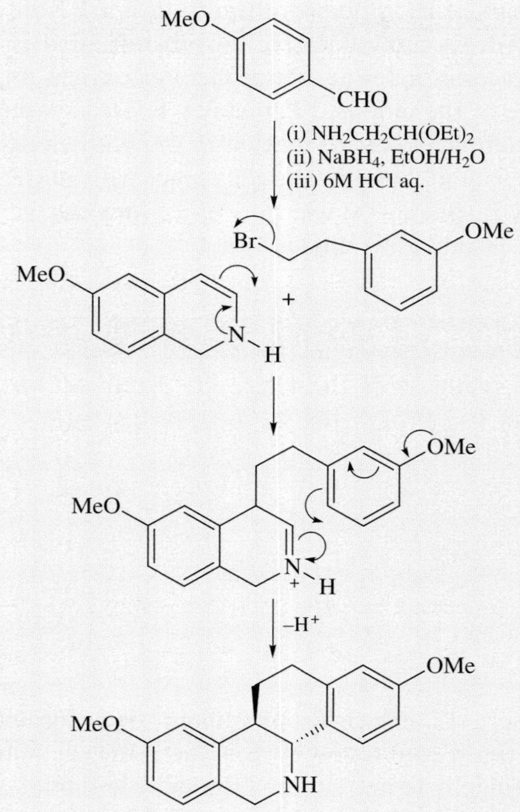

Scheme 3.21

Summary of Key Points

1. Quinoline is attacked by electrophiles at the N atom and in acidic media in the benzenoid ring (the substrate here may be the quinolinium cation).

2. Nucleophiles may react with quinoline at C-2 and C-4; for iso-quinoline, nucleophilic attack occurs at C-1. Such reactions are enhanced if there is a good leaving group at these positions.

3. Both 2- and 4-methylquinolines are similar in reactivity to 2- and 4-methylpyridines and, in general, many of the reactions of pyridines are shared by quinolines and also by isoquinolines.

4. The methods used to synthesize quinolines and isoquinolines are varied, but normally start from phenylamines or similar compounds with an intact benzene ring.

Problems

3.1. Draw a mechanism for the rearrangement of quinoline-8-sulfonic acid into its 6-isomer.

3.2. How might quinoline be converted into quinoline-2-carboxylic acid *via* a Reissert compound and hydrolysis?

3.3. Provide synthetic routes to the quinoline derivatives (a)–(c) from simple benzenes and carbonyl compounds.

(a)

(b) CO_2Et

(c)

Me—   Me   Me

3.4. How could you synthesize 1-methylisoquinoline from benzaldehyde and suitable reagents *via* a multi-step approach?

Further Reading

G. Jones (ed.), *Quinolines,* Wiley-Interscience, Chichester, 1982.

V. T. Andriole (ed.), *The Quinolones*, Academic Press, London, 1988.

C. D. Johnson, *Bicyclic Compounds containing a Pyridine Ring; Quinoline and its Derivatives*, in *Rodd's Chemistry of Carbon Compounds*, 2nd edn., 2nd suppl., vol. F/G, chap. 26, ed. M. Sainsbury, Elsevier, Amsterdam, 1998.

F. G. Kathawala, G. M. Coppola and H. F. Schuster, *Isoquinolines, Parts 1-3*, in *Chemistry of Heterocyclic Compounds*, vol. 31, ed. A. Weissberger and E. C. Taylor, Wiley, New York, 1990.

L. Hazai, *3(2H)-Isoquinolones and their Saturated Derivatives*, in *Adv. Heterocycl. Chem.*, 1991, **52**, 155.

M. Balasubramanian and E. F. V. Scriven, *Isoquinoline and its Derivatives*, in *Rodd's Chemistry of Carbon Compounds*, 2nd edn., 2nd suppl., vol. F/G, chap. 27, ed. M. Sainsbury, Elsevier, Amsterdam, 1998.

4

Pyrylium Salts, Pyrans and Pyrones

Aims

By the end of this chapter you should understand:

- The methods by which the parent ring systems can be synthesized
- The reactions that the various heterocycles undergo with nucleophiles and electrophiles

4.1   Introduction

As oxygen is divalent, no strict equivalent of benzene exists, although the pyrylium cation does achieve aromaticity (see Chapter 1). Both 2*H*-pyrans and 4*H*-pyrans are known, but are encountered more frequently as their carbonyl analogues pyran-2-one and pyran-4-one (see Box 4.1). In addition, reduced forms such as 3,4-dihydro-2*H*-pyran and 3,4,5,6-

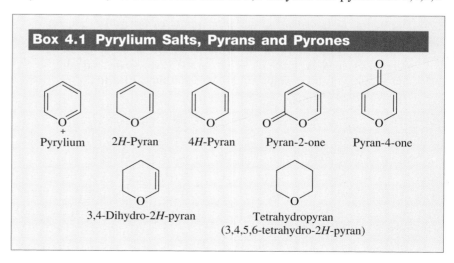

Box 4.1 Pyrylium Salts, Pyrans and Pyrones

Pyrylium   2*H*-Pyran   4*H*-Pyran   Pyran-2-one   Pyran-4-one

3,4-Dihydro-2*H*-pyran

Tetrahydropyran
(3,4,5,6-tetrahydro-2*H*-pyran)

2*H*-tetrahydropyran, the latter better known simply as **tetrahydropyran (THP)**, are valuable laboratory chemicals: 3,4-dihydro-2*H*-pyran is used as a reagent to protect alcohols and THP is a non-protic solvent.

4.2   Pyrylium Salts

4.2.1   Structure and Reactions

Although **pyrylium cations** in combination with an anion of a strong acid are stable, the presence of a formally charged oxygen atom renders them susceptible to reactions with nucleophiles and the valence bond description indicates that C-2(6) and C-4 are the potential targets for attack (Scheme 4.1).

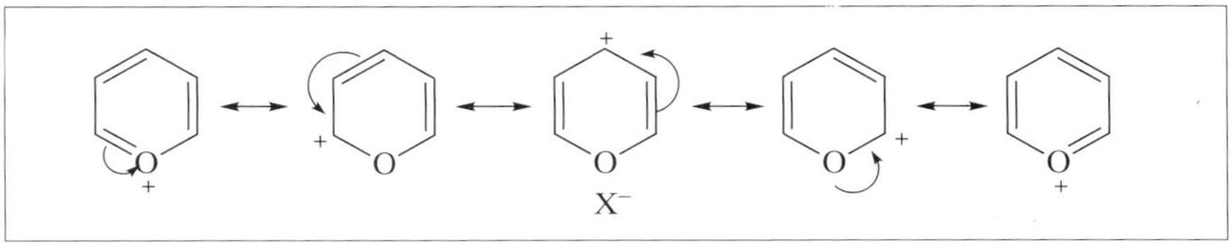

Scheme 4.1

The products of nucleophilic addition are pyrans that frequently ring open and then recyclize to other heterocycles (Scheme 4.2). For this reason, pyrylium salts can be valuable starting materials for a variety of other compounds.

Scheme 4.2 Some nucleophilic reactions of 1,3,5-trimethylpyrylium salts

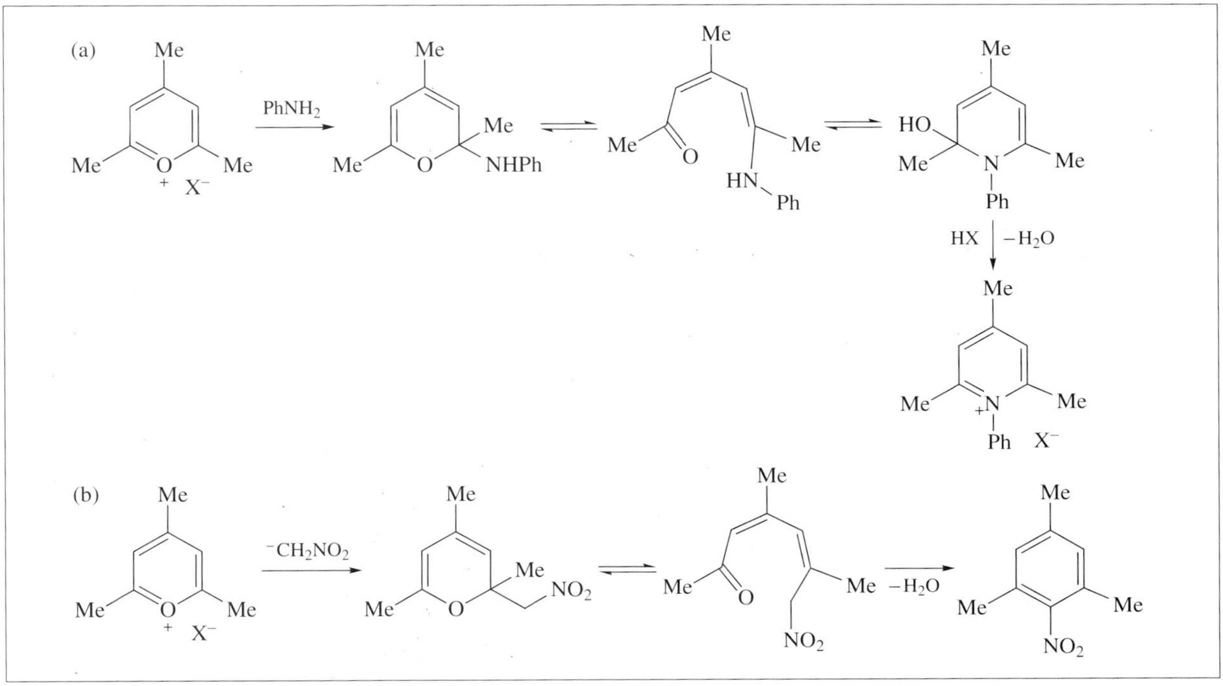

Worked Problem 4.1

Q What sequence of steps takes place when piperidine reacts with a 2,4,6-trimethylpyrylium salt to form *N*-(3,5-dimethylphenyl)-piperidine?

A Piperidine acts both as a nucleophile and as a base. First it combines with the pyrylium cation at C-2, forming an adduct which then ring opens, deprotonates and tautomerizes to an enamino ketone (Scheme 4.3). This product cyclizes through an intramolecular reaction between the enamine unit and the carbonyl group, followed by dehydration to form the phenylpiperidine.

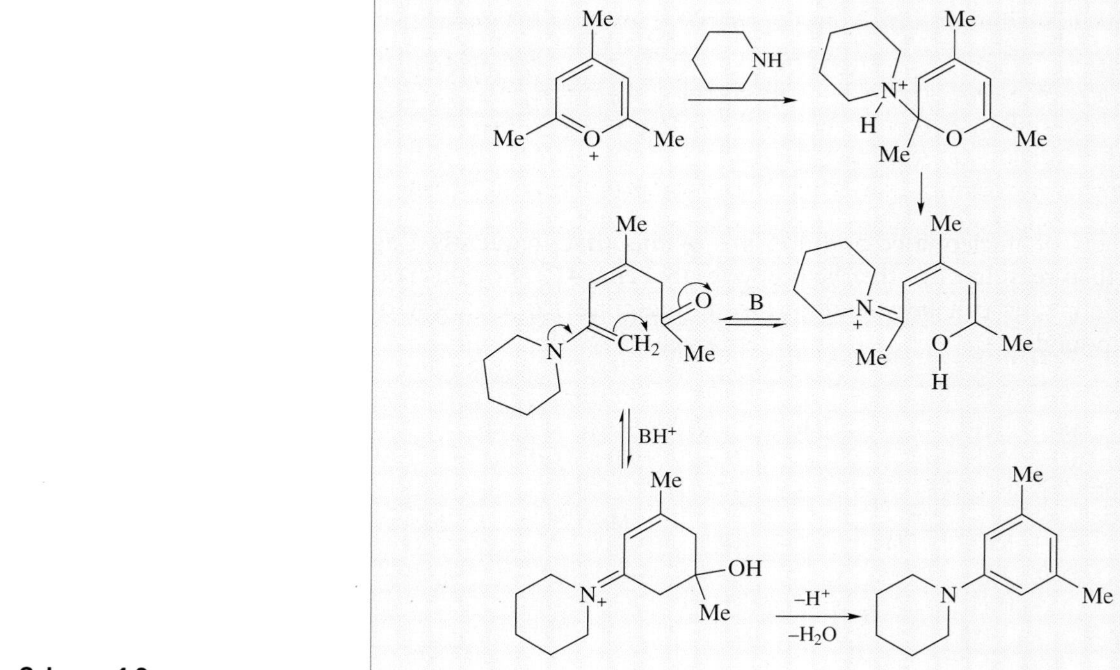

Scheme 4.3

4.2.2 Synthesis

A general method for the synthesis of pyrylium salts is the cyclodehydration of 1,5-dicarbonylalkanes (Scheme 4.4). Acetic anhydride is commonly used as both solvent and reagent, but since the initial product is a 4*H*-pyran, an oxidant such as the **triphenylmethyl (trityl) cation** in the form of triphenylmethyl hexachloroantimonate is added (Ph_3C^+ + [H⁻] → Ph_3CH). In certain cases, however, it is advantageous to isolate the pyran and to oxidize it in a separate step.

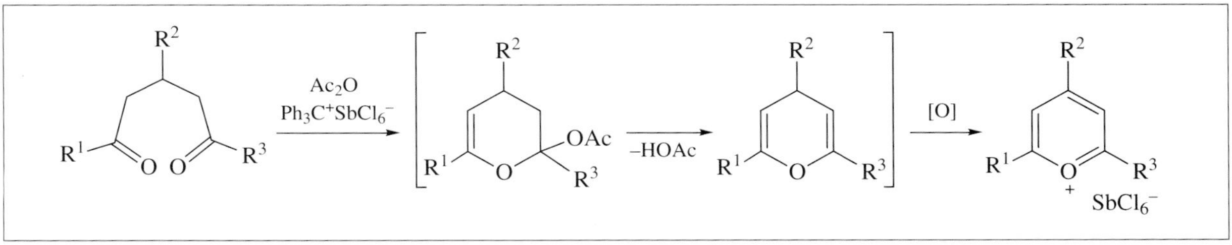

Scheme 4.4

4.3 Pyran-2-ones (α-Pyrones)

4.3.1 Molecular Structure and Electrophilic Reactions

Although the valence bond description of pyran-2-one could include a zwitterionic contribution involving the carbonyl group, thus making the ring aromatic, there is little evidence to support this. For example, the IR carbonyl frequency (1740 cm^{-1}) is typical of an unsaturated lactone, and the chemical shifts of the ring protons in the ^{1}H NMR spectrum indicate that there is no ring current.

Although some electrophilic reagents cause substitution in the ring at C-3 and C-5, very reactive compounds, such as trimethyloxonium tetrafluoroborate (Me$_3$O$^+$BF$_4^-$, Meerwein's reagent) and nitronium tetrafluoroborate (NO$_2^+$BF$_4^-$), attack the carbonyl oxygen atom first and give pyrylium salts (Scheme 4.5). In the case of nitronium tetrafluoroborate the product rearranges into 5-nitropyran-2-one.

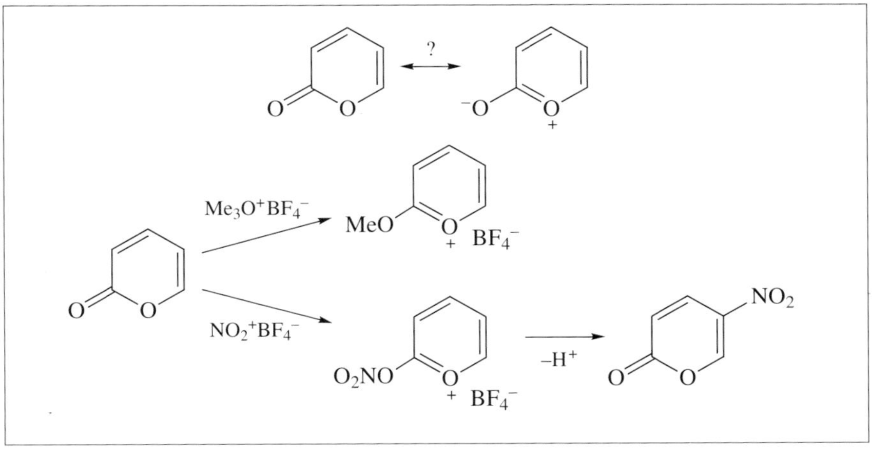

Scheme 4.5

4.3.2 Nucleophilic Attack

The lactonic property of pyran-2-ones is apparent in their reactions with nucleophiles, which normally lead to ring opening. However, this may not be simply direct attack of the reagent at the carbonyl carbon,

The concept of '**hardness and softness**' was first used to provide a non-quantitative assessment of acid and base strengths. It can be extended to characterize nucleophiles and electrophiles. The term 'soft' implies low electronegativity and high polarizability (holds on to electrons loosely), and 'hard' indicates high electronegativity and low polarizablity (holds on to electrons tightly).

as conjugate addition at C-6 is often observed. When it is warmed with aqueous ammonia, pyran-2-one, for example, affords pyridin-2-one, through nucleophilic attack at C-6 followed by recyclization of the initial acyclic intermediate (Scheme 4.6). With the 'harder' reagent methylmagnesium iodide, the reaction seems to proceed by addition to the carbonyl group as the first step. Thereafter the ring opens, and a second equivalent of the Grignard reagent reacts at either of the two potential carbonyl groups in the intermediate. Protonation and dehydration at the end of the process then leads to a ketone and an aldehyde.

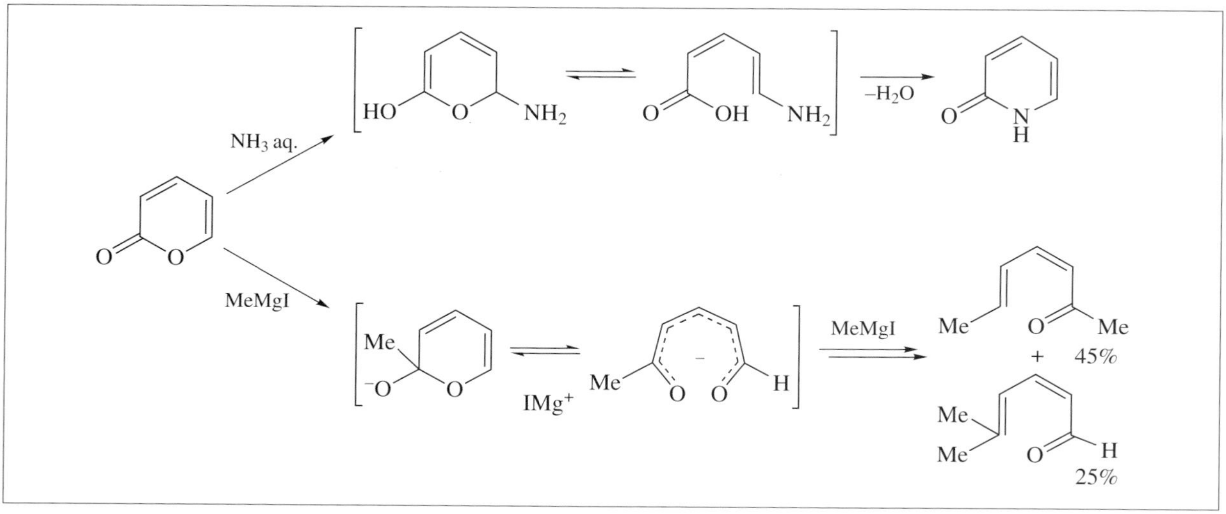

Scheme 4.6

The non-aromatic nature of the pyran-2-one ring is evident in its behaviour as the diene component in **Diels–Alder cycloadditions**. With **dimethyl acetylenedicarboxylate** (**DMAD**, dimethyl but-2-ynedicarboxylate), for example, it gives an adduct that spontaneously eliminates carbon dioxide to yield dimethyl phthalate (dimethyl benzene-1,2-dicarboxylate) (Scheme 4.7)

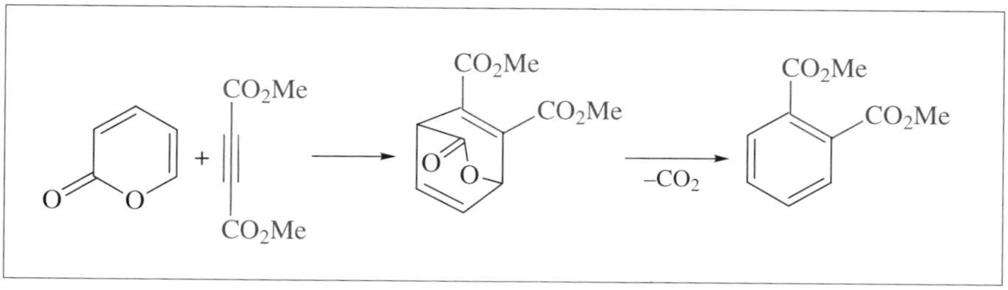

Scheme 4.7

4.3.3 Synthesis

In the **von Pechmann synthesis**, originally used to prepare pyran-2-one-5-carboxylic acid (coumalic acid), 2-hydroxybutane-1,4-dioic acid is converted into 3-oxopropanoic acid by treatment with oleum (sulfur trioxide and conc. sulfuric acid), and this product then self-condenses in the acidic medium to give pyran-2-one-5-carboxylic acid (Scheme 4.8a). Pyran-2-one is obtained by decarboxylating coumalic acid by heating it over copper at 650 °C.

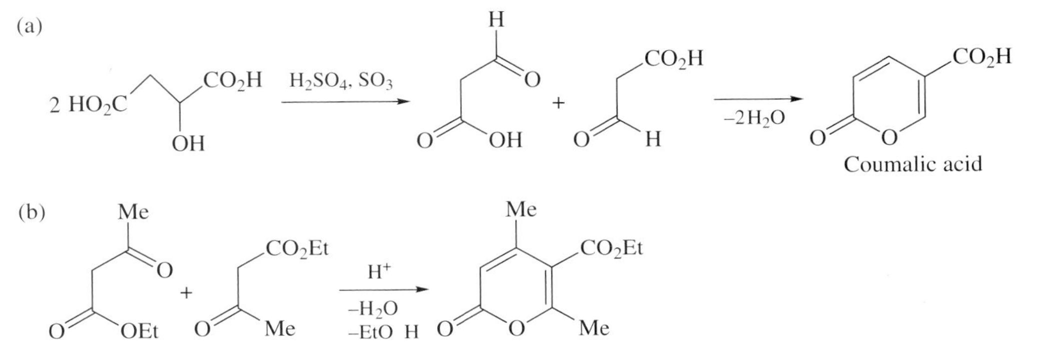

The von Pechmann procedure also works with keto esters. Ethyl acetoacetate (ethyl 3-oxobutanoate), for example, self-condenses readily in the presence of acid to form ethyl 4,6-dimethylpyran-2-one-5-carboxylate (Scheme 4.8b).

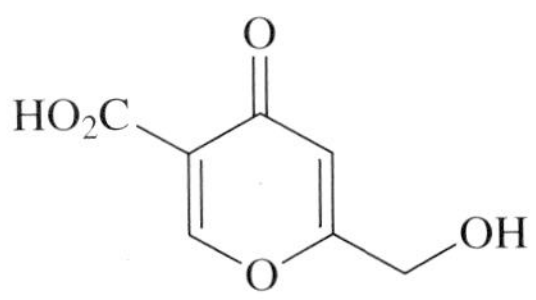

Scheme 4.8

4.4 Pyran-4-ones (γ-Pyrones)

4.4.1 Occurrence and Reactions

Some representatives of this group occur as natural products, including **kojic acid** (**4.1**), which can be isolated from the *Aspergillus* mould used in the manufacture of sake from rice. Like pyran-2-ones, these compounds show little aromatic character and are best considered as vinylogous lactones (Scheme 4.9), rather than as unsaturated ketones.

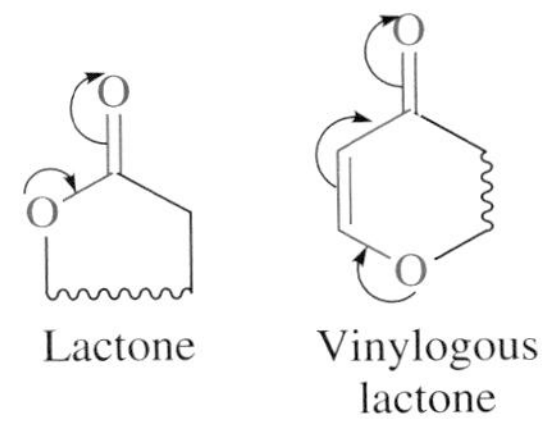

Reactions with electrophiles occur more readily at the carbonyl oxygen atom than is the case for pyran-2-ones. Hydrogen chloride, for example, yields aromatic 4-hydroxypyrylium chlorides (**4.2**) when reacted with 2,6-disubstituted pyran-4-ones.

Nucleophilic attack follows a similar pattern to that encountered with pyran-2-ones; thus hydroxyl ion attacks in conjugate fashion at C-2. In

4.1 Kojic acid

The word **vinylogous** refers to the presence of a CH=CH ('vinyl') unit between two atoms or groups, an O and C=O in this case, relaying the properties of one to the other.

Lactone   Vinylogous
                   lactone

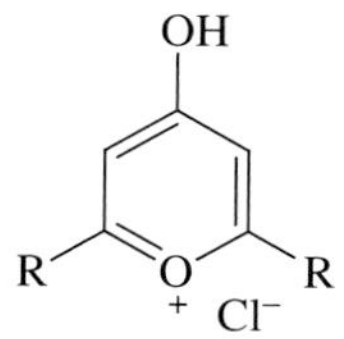

4.2

the case of 2,6-dimethylpyran-4-one, this causes ring opening and recyclization to 1,3-dihydroxy-5-methylbenzene (Scheme 4.10a).

Grignard reagents add directly at C-4; here subsequent ring opening is not an option, but when treated with strong acid the adducts are converted into pyrylium salts (Scheme 4.10b).

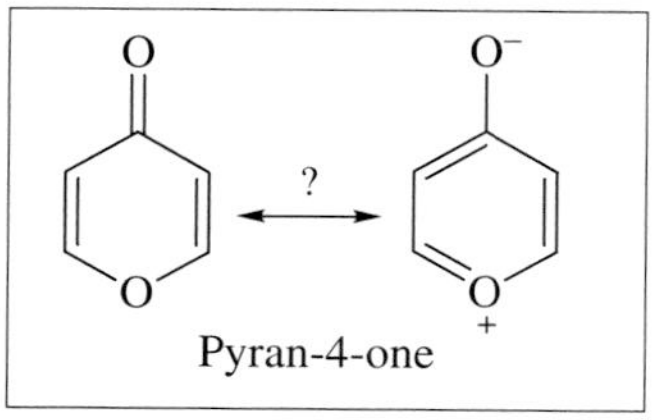

4.4.2 Synthesis

The acid cyclization of 1,3,5-tricarbonyl compounds (the reverse of alkaline hydrolysis) is a common route to pyran-4-ones (Scheme 4.11a). An alternative procedure is the cyclization of alkynyl diketones in the presence of aqueous sulfuric acid (Scheme 4.11b). In this case the substituents R^1 and R^2 can be alkyl or aryl groups.

Pyran-4-one

Scheme 4.9

Scheme 4.10

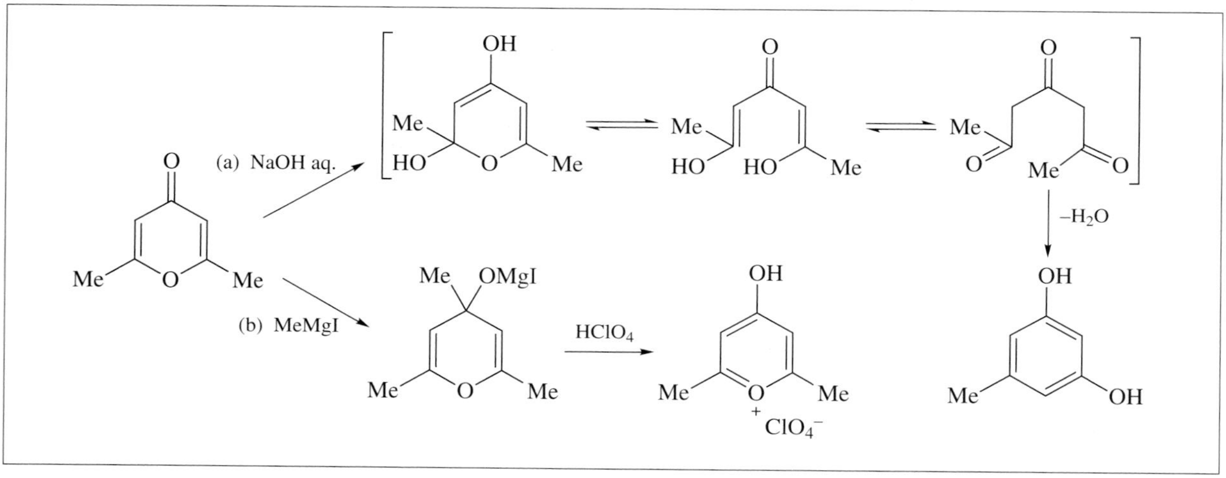

Scheme 4.11

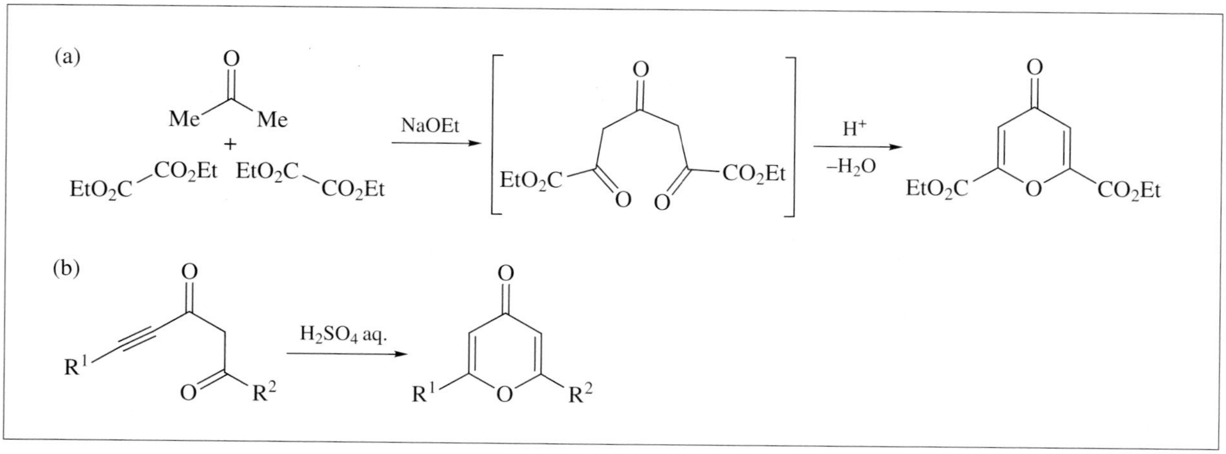

4.5   Reduced Pyrans

Tetrahydropyran can be prepared by heating 1,5-dihydroxypentane at 190–210 °C in the presence of butyltin trichloride (Scheme 4.12a). When the reaction is carried out using a cobalt catalyst [prepared by heating cobalt(II) oxalate (ethanedioate) under hydrogen at 600 °C] instead of the tin halide, the product is 3,4-dihydro-2H-pyran (Scheme 4.12b).

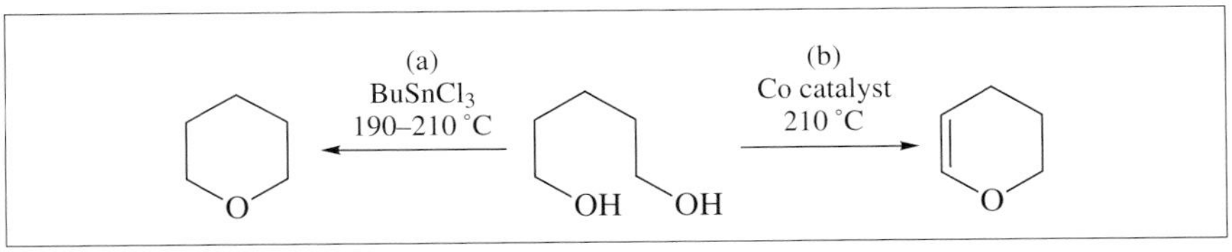

Scheme 4.12

When reacted with an alcohol in the presence of a catalytic amount of a strong acid, 3,4-dihydro-2H-pyran yields a base-stable 2-alkoxytetra-hydropyran (a cyclic acetal) (Scheme 4.13). The reaction is reversed if the acetal is treated with aqueous acid (see Section 3.3.4, page 54), so that this provides a simple way of protecting alcohols in syntheses where a strong base might otherwise deprotonate them. The conformational preferences of 2-alkoxytetrahydopyrans, mediated by the anomeric effect, were commented upon earlier (Section 1.5.3).

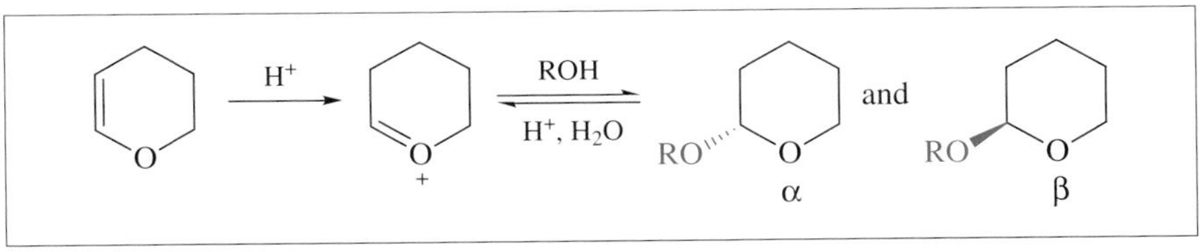

Scheme 4.13

4.6   Saccharides and Carbohydrates

A major group of tetrahydropyrans is represented by the pyranose mono-saccharides. These polyhydroxylated compounds, together with isomers based upon tetrahydrofuran (furanoses), are a very important class of natural products and also form the building blocks from which poly-saccharides are constructed. The 2-position or the anomeric position in the monosaccharides is the usual linkage site. Sugars in the α-series have an oxygen substituent at the anomeric position *below* the plane of the ring, whereas those with an oxygen substituent *above* the ring plane at this site are members of the β-series.

Sucrose (**4.5**, common sugar) is an acetal formed from glucose (**4.3**) and fructose (**4.4**); ignoring terms which indicate the absolute stereo-chemistry, a more complete name for sucrose is 2α-glucopyranose-2β-fructofuranose.

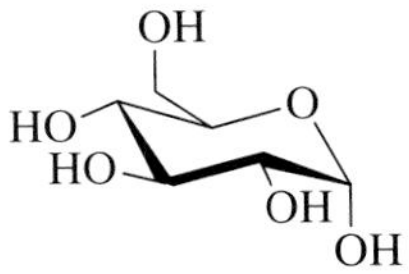

4.3 α-Glucose
(glucopyranose)

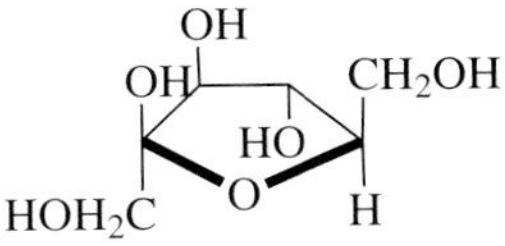

4.4 β-Fructose
(fructofuranose)

The chemistry of sugars (**carbohydrates**) is a major subject area, covered by many specialist texts. Note: monosaccharides can exist as open-chain or as cyclic forms, where the oxygen atom forms part of the ring. Those with six-atom rings are called **pyranoses**, and those with five-atom rings are described as **furanoses**.

4.5 Sucrose

Summary of Key Points

1. Pyrones exist as such; there is little evidence that the ring is aromatic.

2. Nucleophiles attack either the carbonyl group of pyran-4-ones or at C-2. In pyran-2-ones, conjugative addition at C-6 can occur.

3. 'Hard' electrophiles attack pyran-2-ones at the carbonyl oxygen atom.

4. Pyran-2-ones are effective dienes in cycloaddition reactions.

Problems

1. Suggest a mechanism for the conversion of pyridinium-*N*-sulfonate into pyrylium perchlorate [chlorate(VII)] in two steps, the first being the action of aqueous sodium hydroxide, and the second treatment with perchloric acid in diethyl ether and methanol.

2. Suggest a synthetic route to the tricyclic pyrylium salt **4.6** from 2-formylcyclohexanone and perchloric acid plus a simple ketone.

3. How might the 2-pyrone **4.7** be converted into the arene **4.8**?

4. Suggest a synthesis of chelidonic acid **4.9** from diethyl oxalate [(CO$_2$Et)$_2$] and acetone (propanone).

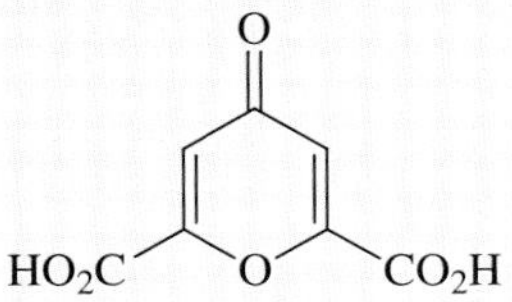

4.9

Further Reading

K. Dimroth and K. H. Wolf, *Aromatic Compounds from Pyrylium Salts*, in *Newer Methods of Preparative Organic Chemistry*, vol. 3, Academic Press, New York, 1964.

A. T. Balaban, W. Schroth and G. W. Fischer, *Pyrylium Salts. Part 1, Syntheses*, in *Adv. Heterocycl. Chem.*, 1969, **10**, 241.

A. T. Balaban, A. Dinculescu, G. N. Dorofeeko, G. W. Fischer, A. V. Koblik, V. V. Metzheritskii and W. Schroth, *Pyrylium Salts. Synthesis, Reactions and Physical Properties*, in *Adv. Heterocycl. Chem., Suppl. 2*, 1982.

5

Benzopyrylium Salts, Coumarins, Chromones, Flavonoids and Related Compounds

Aims

At the end of this chapter you should understand:

- The methods by which the parent ring systems can be synthesized
- Some differences between the chemical reactions of coumarins and chromones

5.1   Structural Types and Nomenclature

There are numerous examples of benzopyrylium salts, benzopyrans and benzopyranones, and frequently they have trivial names that reflect their long history (see Box 5.1). Many are natural products, and frequently these compounds contain hydroxy or alkoxy groups (sometimes in the form of a sugar residue). Polyhydroxylated natural products based upon 2-phenylbenzopyrylium (flavylium) salts and with ether linkages to sugars are called anthocyanins, whereas without their sugars they are known as anthocyanidins.

Anthocyanins, in association with other compounds, such as flavones, are responsible for the colour of certain flowers. An anthocyanin found in rose petals is cyanin; it can be isolated as its chloride. The corresponding anthocyanidin, cyanidin, exists as the pentahydroxy salt in acidic media, but as the pH increases it gives first a quinone and then an anion. Each of these forms has a different colour (see Scheme 5.1).

Other natural products in the group are built up by ring fusions of several types of ring systems. They include the garden insecticide rotenone (sometimes sold in the crude form as derris powder, the pulverized bark of the plant *Derris elliptica*).

The general term **glycoside** is applied to natural products bonded to one or more sugars through ether groups. The corresponding term **aglycone** is used for the hydroxylated derivatives, where the sugar residues are no longer present. Anthocyanins are thus glycosides, whereas anthocyanidins are aglycones.

Box 5.1 Benzopyrylium Salts, Coumarins, Chromones and Flavonoids

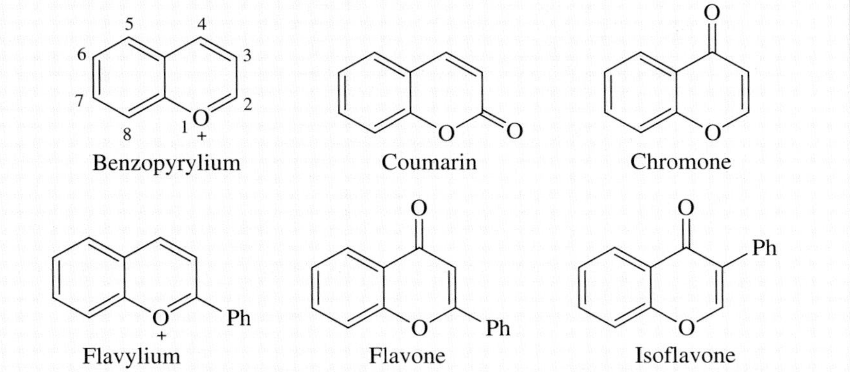

Benzopyrylium

Coumarin

Chromone

Flavylium

Flavone

Isoflavone

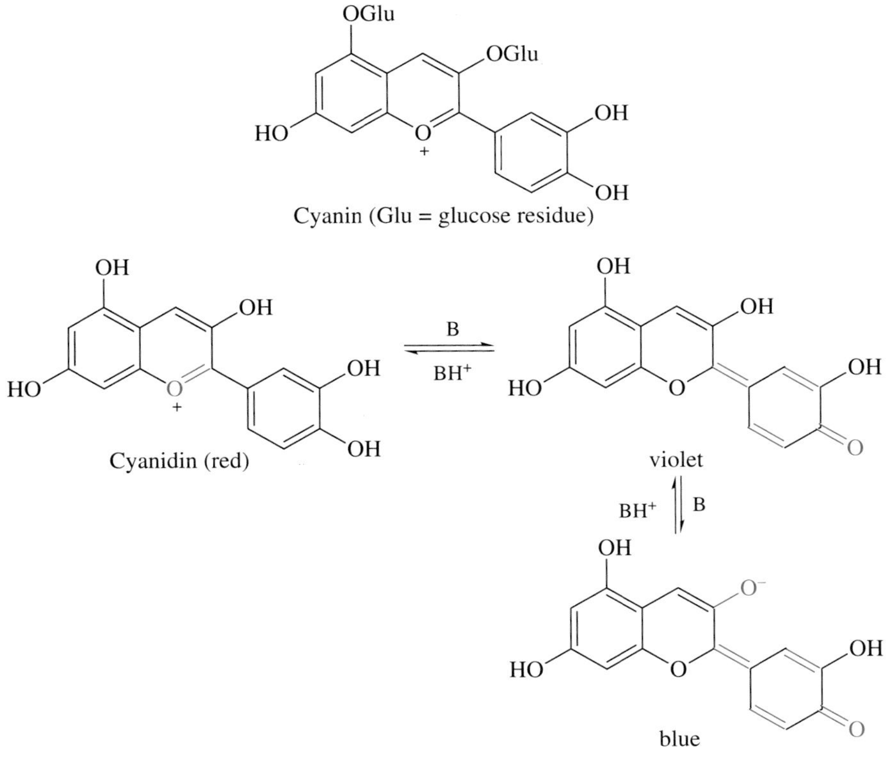

Cyanin (Glu = glucose residue)

Cyanidin (red)

violet

blue

Scheme 5.1

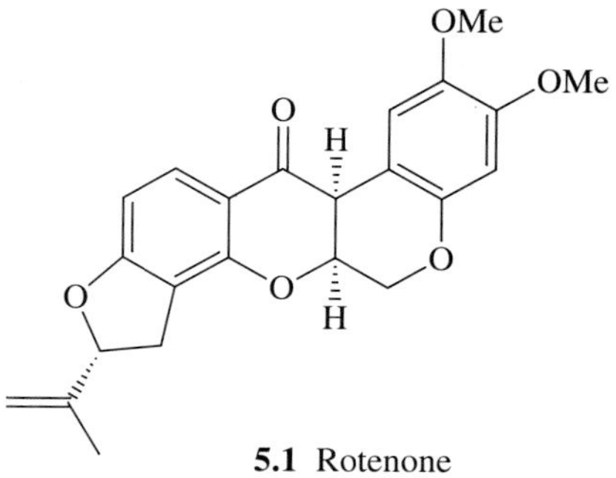

5.1 Rotenone

5.2   Coumarins

5.2.1 Introduction

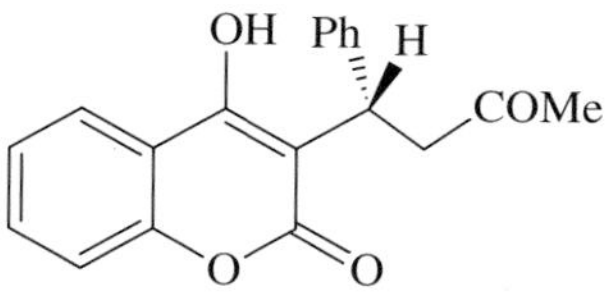

5.2 Warfarin

Hydroxylated coumarins are present in grass and contribute to the smell of newly cut hay; others have pharmaceutical and rodenticidal activities. The compound warfarin (**5.2**) was developed to kill rats, but is now often used as a blood anti-coagulant in human patients.

5.2.2 Reactions

Resonance within the unsaturated lactone unit of coumarin gives a strong hint as to its likely reactivity. Thus, the oxygen atom of the carbonyl group receives electron density both *via* the enone chromophore and from the internal resonance of the lactone group (Scheme 5.2).

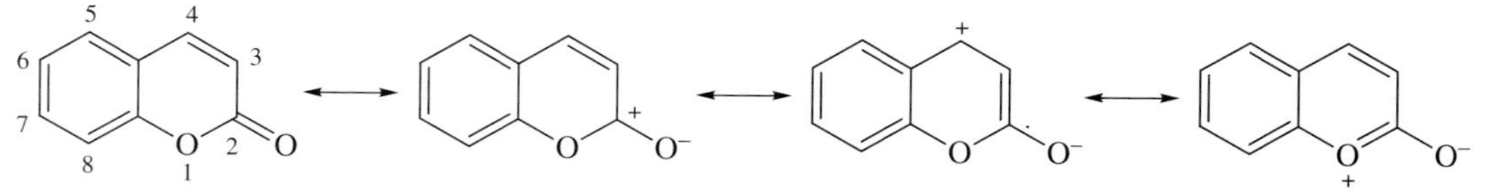

Scheme 5.2

Nucleophilic addition occurs mainly at the carbon atom of the carbonyl group, causing ring opening. Similarly, electrophilic reagents containing an element capable of forming a strong bond to oxygen (oxophiles) bind to the oxygen atom of the carbonyl group; thus silanes, for example, give 2-(*O*-silyl)benzopyrylium salts.

Other less oxophilic electrophiles give C-6 substituted coumarins, but it is unclear whether the substrate for such reactions is the free coumarin or a cation formed by protonation or bonded by a Lewis acid at the carbonyl oxygen. Some typical reactions are shown in Scheme 5.3.

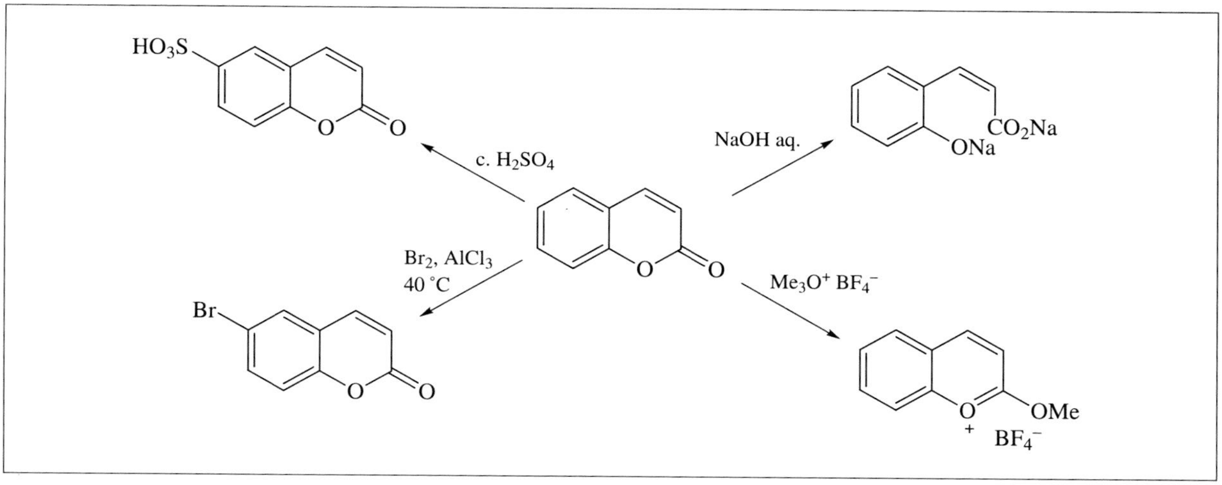

Scheme 5.3

Worked Problem 5.1

Q Suggest a possible mechanism for the conversion of coumarin into 3-bromocoumarin by the action of bromine and then pyridine.

A In the absence of a Lewis acid, bromine adds across the 3,4-double bond of coumarin to give 3,4-dibromo-3,4-dihydrocoumarin. In the presence of pyridine a dehydrobromination reaction takes place, leading to 3-bromocoumarin as the favoured product (Scheme 5.4).

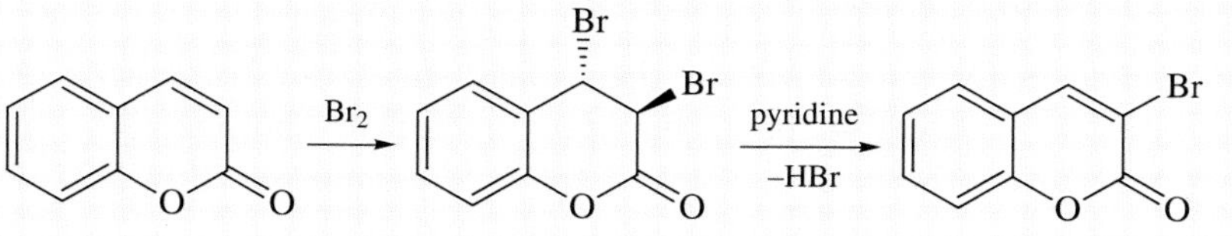

Scheme 5.4

Assuming that the dehydrobromination step follows an E1 mechanism, then the chemoselectivity is accounted for by the participation of the better carbocation (*i.e.* a benzylic cation where the + charge is conjugated with the benzene nucleus). The alternative cation places the positive charge next to the carbonyl group:

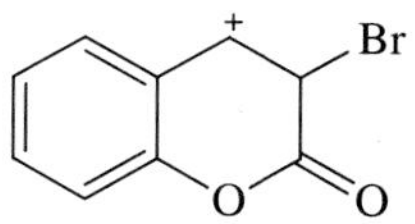

benzylic cation

versus

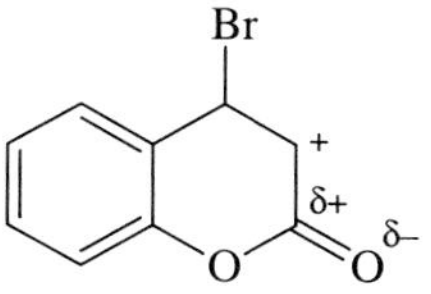

5.2.3 Synthesis

One approach is to use a 2-hydroxybenzaldehyde to form all but two atoms of the molecule. The remaining atoms are supplied by malonic acid (propane-1,3-dioic acid), which combines with the aldehyde in a Knoevenagel condensation step, before cyclization (lactonization) and decarboxylation occur (Scheme 5.5).

4-Methylcoumarins bearing hydroxy and other electron-donating groups can be synthesized from the corresponding phenols by reaction with ethyl acetoacetate in the presence of sulfuric acid. Hydrolysis of the ester group in the product then allows the lactone ring of the coumarin to form (Scheme 5.6).

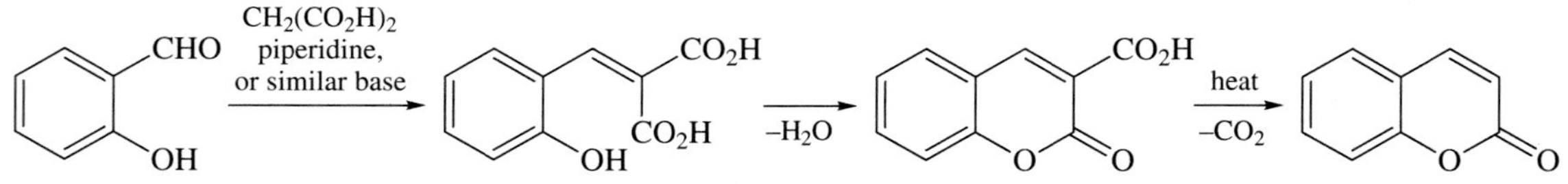

Scheme 5.5

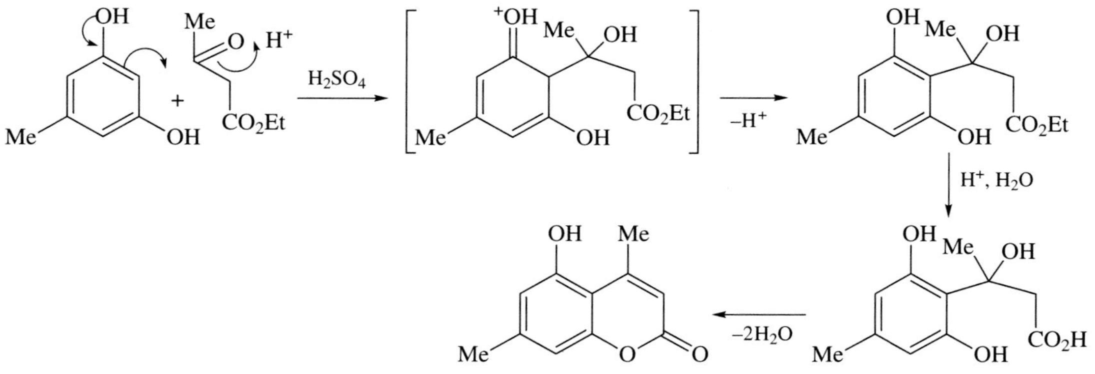

Scheme 5.6

5.3   Chromones (Benzopyran-4-ones)

5.3.1   Structure and Reactions

Chromones differ marginally in their chemistry from coumarins (ben-zopyran-2-ones) because the carbonyl group is now conjugated with the oxygen atom *via* the double bond of the heterocycle (see Box 5.2). This conjugation does not involve the benzene ring.

Box 5.2 Chromone Conjugation

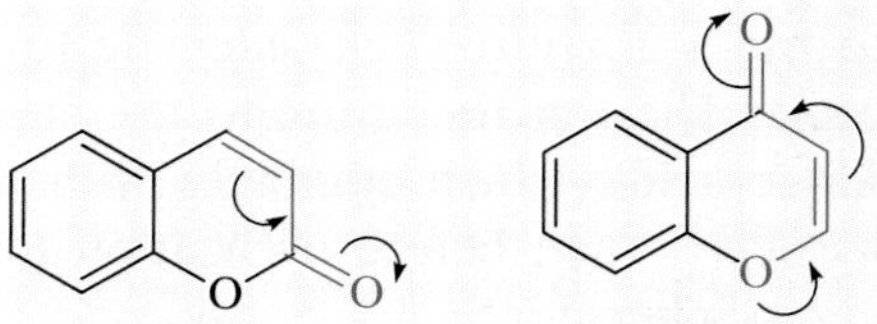

Coumarin   Chromone

As a result, chromones are rather more basic, and strong acids read-ily protonate the carbonyl oxygen atom, forming crystalline benzopyrylium salts. Once protonated, the molecule should be resistant to further electrophilic attack, but with fuming nitric and concentrated sulfuric acid at between 0 °C and room temperature, chromone gives the 6-nitro derivative (Scheme 5.7). Chromone also undergoes a Mannich-type reaction (see Section 7.1.2) with dimethylamine and formaldehyde (methanal) in hydrogen chloride and ethanol; here the product is 3-(*N,N*-dimethylaminomethyl)chromone.

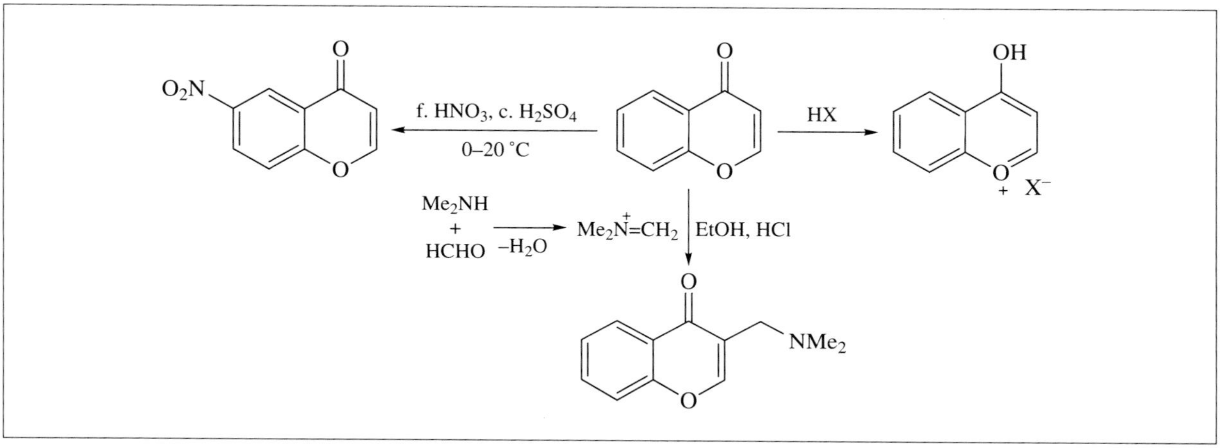

Relatively hard nucleophiles, such as Grignard reagents, may attack at   **Scheme 5.7**
the carbonyl carbon, whereas softer nucleophiles, *e.g.* hydroxide ion, combine at C-2 by conjugative addition, and this may then cause ring opening (Scheme 5.8).

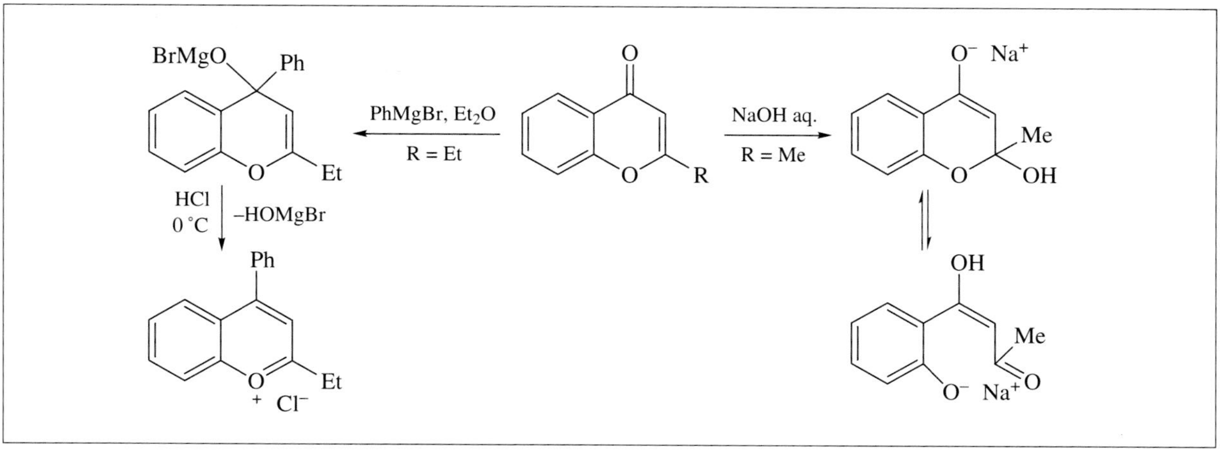

Scheme 5.8

Worked Problem 5.2

Q Why should the Mannich reaction with chromone take place at C-3?

A The Mannich reaction proceeds through the intervention of the N,N-dimethylmethyleniminium cation [$Me_2N^+=CH_2$]. This is insufficiently electrophilic to react with the benzene ring under the mild reaction conditions. Similarly, were the electrophile to react with the carbonyl oxygen atom of the heterocycle, this reaction would be reversible, as an aminomethyl ether is relatively unstable in acidic media. Thus, it seems plausible that the chromone utilizes 'enol or enolate' character to trap the electrophile at C-3, followed by deprotonation of the adduct to reform the chromone ring system (Scheme 5.9).

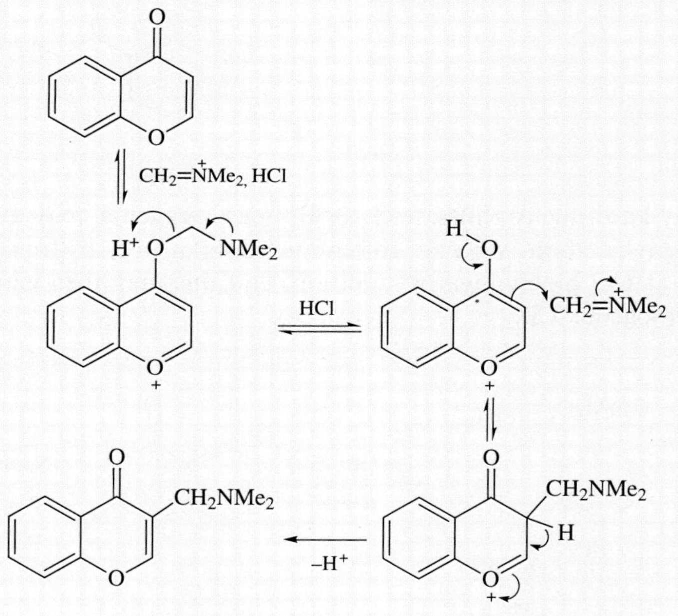

Scheme 5.9

5.3.2 Synthesis

Syntheses of 2- and 3-substituted chromones normally start from 2-hydroxyphenyl ketones. In the first of two examples, a route to flavone is shown in Scheme 5.10 using 2-hydroxyacetophenone (2-hydroxyphenylethanone) and benzoyl chloride as starting materials. Initially, the phenolic group of the acetophenone is *O*-acylated by benzoyl chloride, using pyridine as a base (a Schotten–Baumann-type reaction).

Under these conditions, the *O*-benzoyl derivative immediately enolizes and is *O*-acylated again to yield a dibenzoate. Without isolation, this product is cyclized by treatment with aqueous potassium hydroxide to yield 2-hydroxy-2,3-dihydroflavone. Dehydration to flavone is then effected by the action of glacial acetic acid containing sulfuric acid.

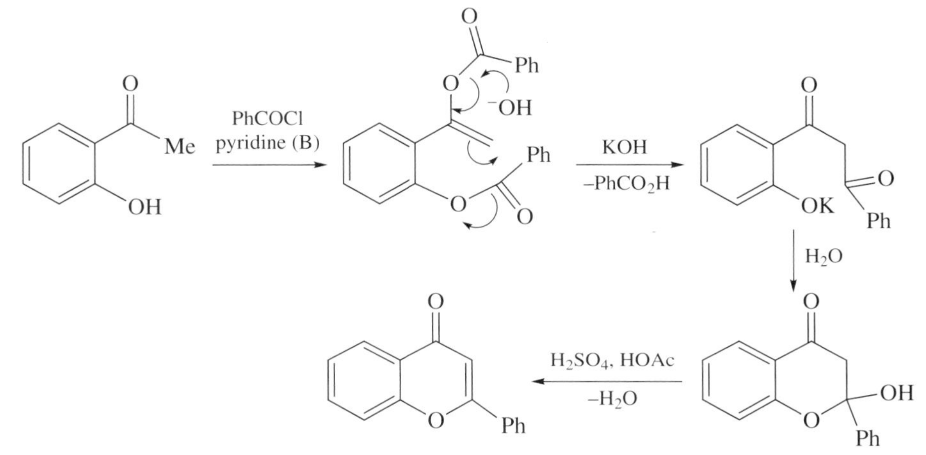

Scheme 5.10

A similar route to 3-substituted chromones and isoflavone (R = Ph) relies upon a **Claisen-like condensation** between the enolate of a 2-hydroxyphenyl ketone and ethyl formate (methanoate) (Scheme 5.11). This affords a 2-hydroxydihydrochromone that, as in the first example, is subjected to an acid-promoted dehydration in the final step.

Dihydrochromones are often called **chromanones**, and dihydroflavones and dihydro-isoflavones are known as **flavanones** and **isoflavanones**, respectively.

Scheme 5.11

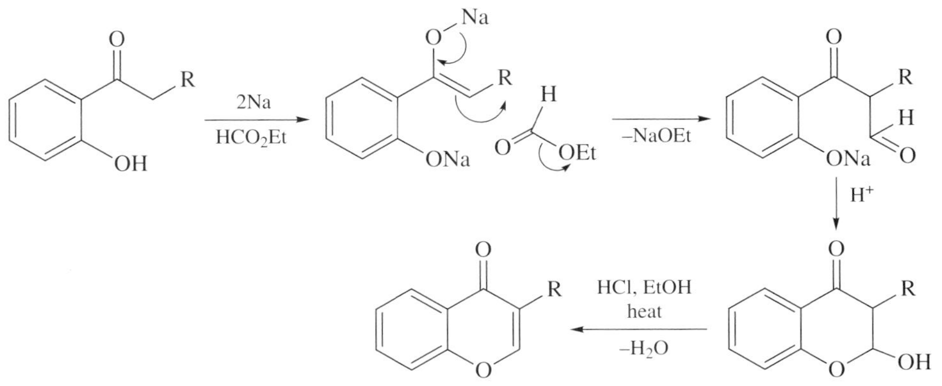

Summary of Key Points

1. The benzopyrylium cation is aromatic in nature, but the heterocyclic ring is readily reduced and reacted with nucleophiles.

2. The presence of a benzene unit fused to a pyrone ring affects the chemistry of both coumarins and chromones, but there are subtle differences in reactivity between the two types of compounds.

3. As for pyrones, the nature of the nucleophile (hard or soft) determines whether reactions occur at the carbonyl group or at the β-atom of the enone system. Electrophiles attack in the benzene ring.

Problems

1. Devise syntheses for (a) 3-methylcoumarin and (b) 3-methylchromone from 2-hydroxybenzaldehyde and ethyl 2-hydroxyphenyl ketone, respectively.

2. Suggest mechanisms for the ring opening of (a) coumarin to 2-hydroxycinnamic acid [3-(2-hydroxyphenyl)propenoic acid] by treatment with aqueous sodium hydroxide, and (b) a mechanism whereby 2-methylchromone is ring opened under the same conditions to 2-(3-oxobutanoyl)phenol (2-$HOC_6H_4COCH_2COMe$).

Further Reading

J. B. Harborne, T. J. Mabry and H. Mabry (eds.), *The Flavonoids*, Chapman Hall, London, 1975.

C. O. Chichester (ed.), *The Chemistry of Plant Pigments*, Academic Press, New York, 1972.

F. M. Dean (ed.), *The Natural Coumarins: Occurrence, Chemistry and Biochemistry*, Wiley, Chichester, 1982.

6
Five-membered Heterocycles containing One Heteroatom: Pyrrole, Furan and Thiophene

Aims

By the end of this chapter you should be able to understand:

- Why pyrrole, furan and thiophene show aromatic character, and the origins of chemo- and regioselectivity in their reactions with electrophiles
- That in the case of pyrrole the formation of the pyrrole anion and the nature of its reactions is of importance, and for furan the reason why cycloaddition reactions occur readily
- How to synthesize the heterocycles, and how simple functional group transformations can be carried out to make derivatives

6.1 Pyrrole

6.1.1 Introduction

Formally, **pyrrole** can be described as azacyclopentadiene, *i.e.* cyclopentadiene in which a CH_2 unit has been replaced by a NH group. However, this translates into a classical structure that does not adequately describe the compound, for pyrrole has aromatic character, even though there are only five atoms in the ring! The aromaticity arises because the nitrogen atom contributes *two* electrons and the four carbons *one electron each* to form a delocalized sextet of π-electrons. In valence bond terms the structure of pyrrole can be represented as a resonance hybrid (Scheme 6.1)

Although resonance causes a partial negative charge to reside on the carbon atoms and a partial positive charge on the nitrogen atom, this polarization is offset by an inductive effect in the opposite direction (nitrogen is more electronegative than carbon!). Overall, the balance is

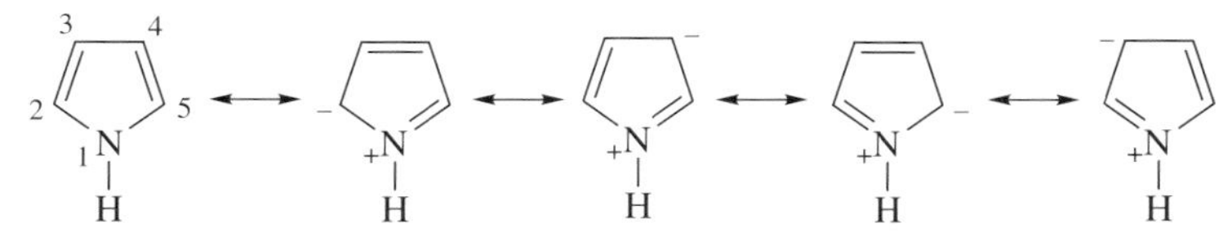

Scheme 6.1

See Section 2.1 for a brief
description of dipole moment.

in favour of the resonance effect and pyrrole has a dipole moment
(1.55–3.0 D) directed *away* from the nitrogen atom (the value of the
dipole moment depends on the solvent). Pyrrole has about half the resonance energy of benzene, and it is only a very weak base (pK_a –3.6).

6.1.2 Electrophilic Substitution

Pyrrole generally reacts with electrophiles (E^+) (including the proton) at
C-2 in preference to C-3. This selectivity can be explained by more exten-
sive delocalization of the positive charge in the σ intermediate for C-2
substitution than is the case in the equivalent intermediate for attack at
C-3 (Scheme 6.2).

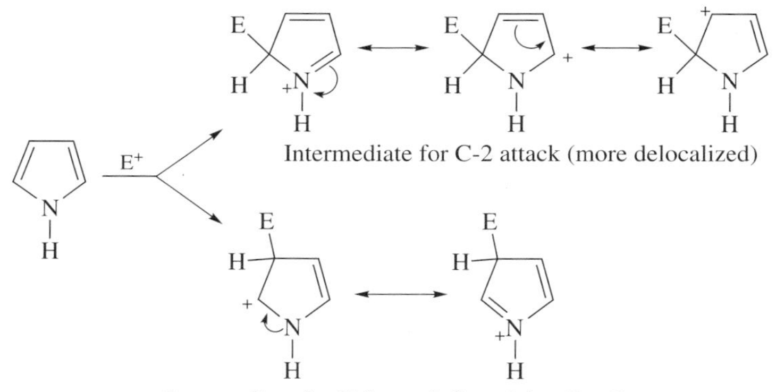

Intermediate for C-2 attack (more delocalized)

Intermediate for C-3 attack (less delocalized)

Scheme 6.2

Although these explanations
represent good 'rules of thumb',
the reactions of some
electrophiles, including the
proton, can be complex; for a full
interpretation, thermodynamic and
kinetic factors also need to be
considered.

Attack at nitrogen is inhibited because no delocalization of charge is
possible in the intermediate (Scheme 6.3). *N*-Alkylation and -acylation
of pyrrole are best carried out indirectly using the anion of pyrrole (see
Section 6.1.3).

Pyrrole is very reactive towards electrophiles. With bromine in ethanol
at 0 °C, for example, it gives 2,3,4,5-tetrabromopyrrole, whereas sulfuryl
chloride (SO_2Cl_2) in diethyl ether at 0 °C yields 2-chloropyrrole.

Typical electrophilic reactions, such as nitration, halogenation with a
Lewis acid (as a 'carrier'), Friedel–Crafts C-alkylation and -acylation,
that work well with benzene, cannot be applied to pyrrole, because heating with strong acids, or a Lewis acid, destroys the heterocycle. However,

milder conditions can sometimes be used. With pyridinium-*N*-sulfonate, for example, a 90% yield of the corresponding pyrrole-2-sulfonic acid is obtained, after acidification (Scheme 6.4).

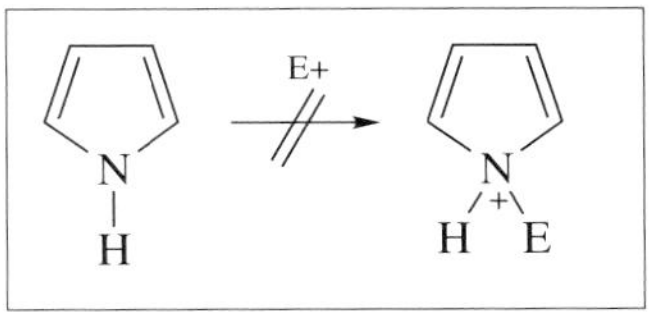

Scheme 6.3

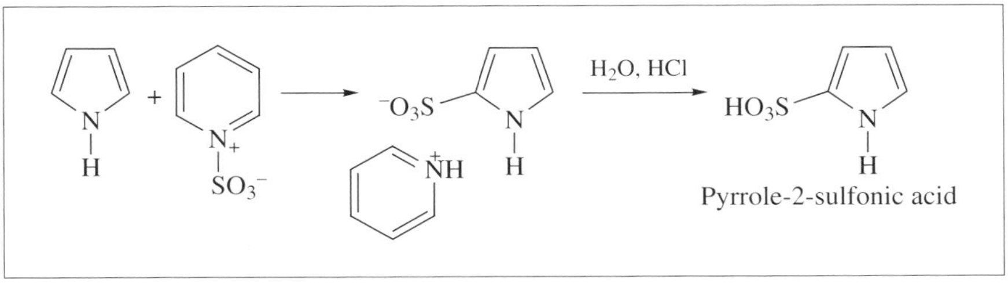

Scheme 6.4

Similarly, the **Vilsmeier reaction** (reagents: $POCl_3$ and *N,N*-dimethylformamide) gives 2-formylpyrrole (Scheme 6.5).

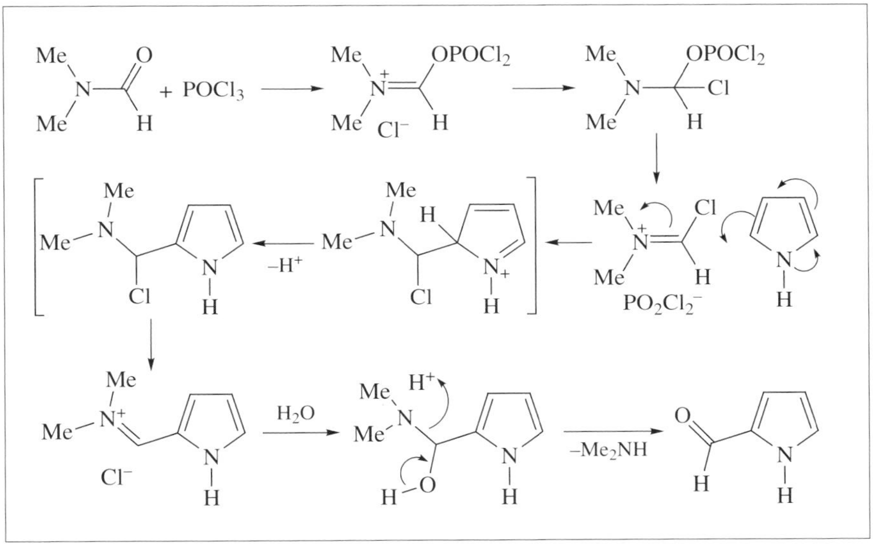

Scheme 6.5

Worked Problem 6.1

Q Explain how pyrrole reacts with $Me_2NCHO/POCl_3$ to give a product that can be further reacted with $MeNH_2$ and then with sodium borohydride in ethanol to give 2-[(*N*-methylamino)-methyl]pyrrole.

A The reaction conditions of the first step are those of the Vilsmeier reaction (see above). With the electrophile $Me_2N^+=CHCl$, pyrrole is attacked at C-2, rather than at C-3, and after hydrolysis 2-

formylpyrrole is obtained. This aldehyde combines next with methylamine first to form an imine that undergoes reduction on treatment with $NaBH_4$ giving 2-[(*N*-methylamino)methyl]pyrrole (Scheme 6.6).

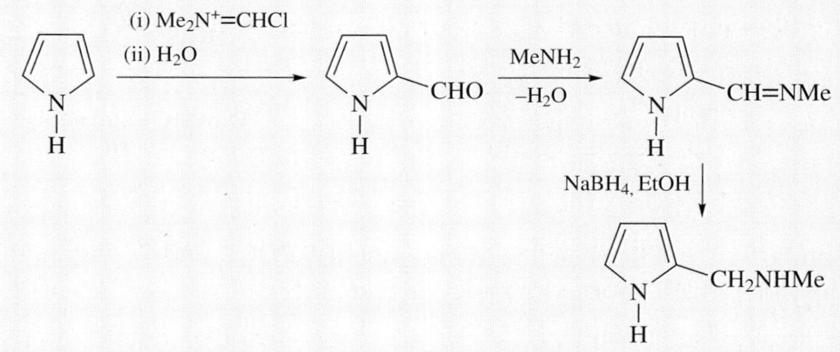

Scheme 6.6

Pyrrole reacts with aldehydes and ketones and an acid catalyst to form resins, probably linear polymers; however, surprisingly, from an entropic point of view, with acetone (propanone) and hydrochloric acid the product is a cyclic tetramer. Possibly the two methyl groups in the developing side chains force it to bend and thus 'bite its own tail' (Scheme 6.7).

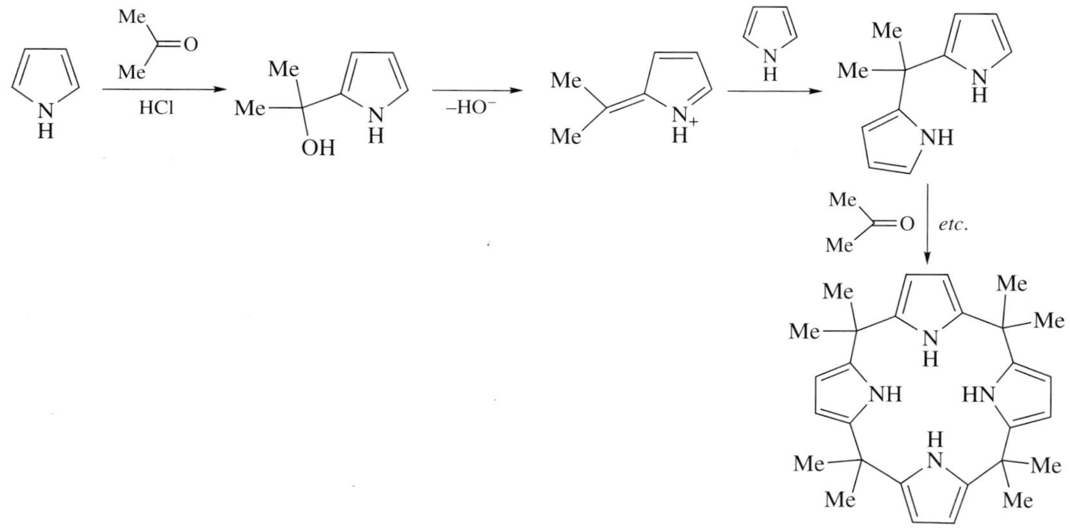

Scheme 6.7

6.1.3 Pyrrole Anion

Although pyrrole is a very weak acid (pK_a 17.7), it can be deprotonated by a strong base, such as butyllithium. Its 'acidity' is much greater than a typical aliphatic secondary amine, say **pyrrolidine** [tetrahydropyrrole, (pK_a *ca.* 27)]. Unlike pyrrole itself, resonance within the pyrrole anion does not involve charge separation (Scheme 6.8).

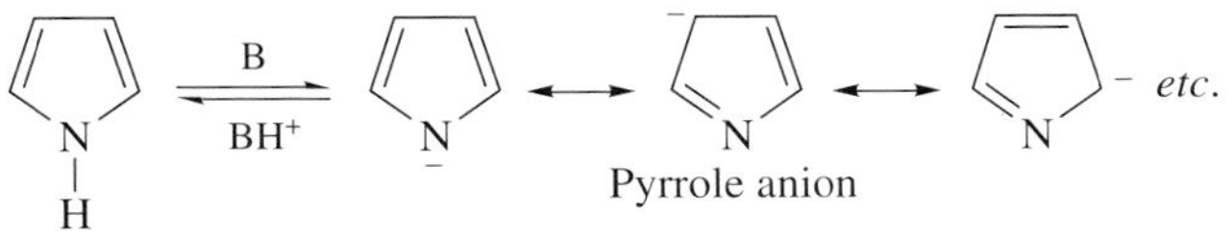

Scheme 6.8

Although pyrrole is much less acidic than phenol (pK_a *ca.* 10), many reactions of the pyrrole anion are reminiscent of those of the phenolate anion. Examples include *N*-alkylation, -acylation and -sulfonylation (Scheme 6.9). For the phenolate anion these reactions occur at oxygen.

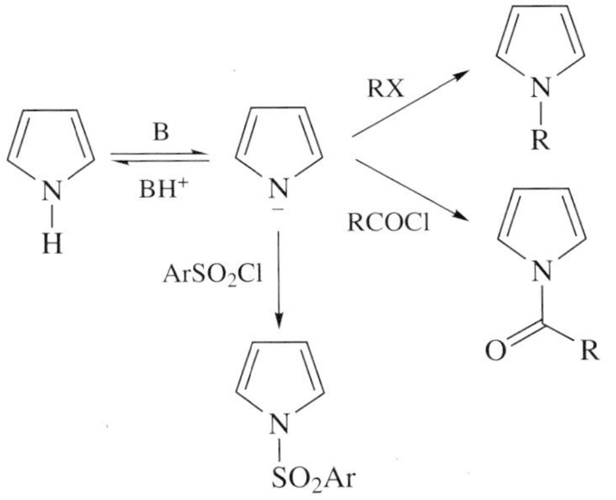

Scheme 6.9

The pyrrole anion also undergoes the **Kolbe carboxylation reaction** with carbon dioxide under pressure in a heated autoclave and gives pyrrole-*N*-carboxylic acid, on acidic work-up (Scheme 6.10).

The **Kolbe reaction** is best known for the conversion of sodium phenolate into salicylic acid (2-hydroxybenzoic acid).

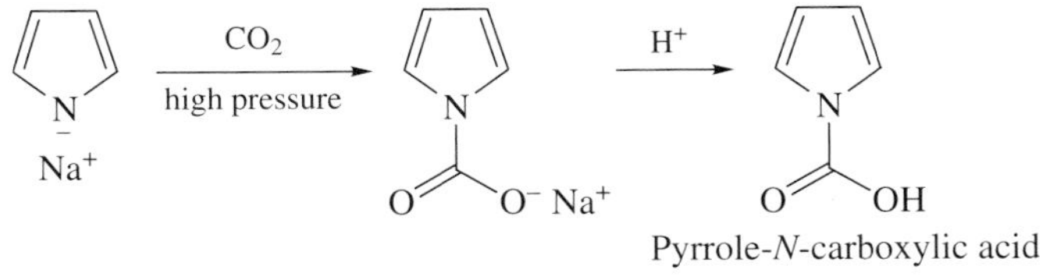

Scheme 6.10

Singlet dichlorocarbene (generated from chloroform by the action of a strong base, such as potassium *tert*-butoxide, KOCMe$_3$), reacts with the anion of pyrrole to give an adduct, which then ring expands to give 3-chloropyridine (Scheme 6.11).

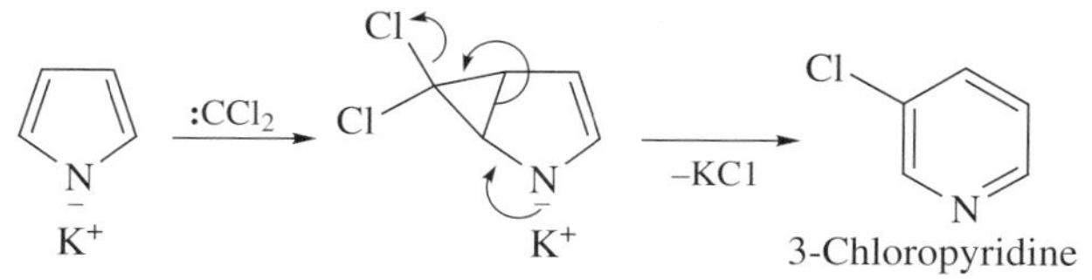

Scheme 6.11

6.1.4 Cycloaddition Reactions

Pyrrole and its simple derivatives do not react easily as dienes. Pyrrole itself only combines with dimethyl acetylenedicarboxylate (DMAD, dimethyl but-2-ynedicarboxylate) under high pressure and then it is by C-2 substitution. However, *N*-acylpyrroles, such as *N*-acetyl- and *N*-(*tert*-butoxycarbonyl)pyrrole, do undergo **Diels–Alder addition reactions**. Here, internal resonance within the acyl group reduces the availability of the lone-pair electrons, formally on nitrogen, to delocalize into the ring, thus making the carbon unit more 'diene-like' (Scheme 6.12).

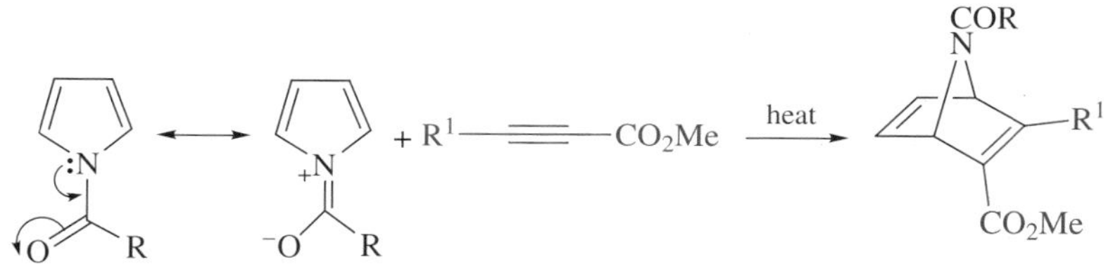

Scheme 6.12

6.1.5 Synthesis

There are two principal routes to pyrroles. One is called the **Paal–Knorr synthesis**, in which pyrroles are formed by the interaction of 1,4-dicarbonyl compounds and ammonia. No intermediates have ever been isolated, so the mechanism shown in Scheme 6.13 is speculative.

A recent adaptation of this type of approach involves the spontaneous intramolecular cyclization of imino ketones, formed by the reduction of nitro ketones through the action of tributylphosphine/diphenyl sulfide (Scheme 6.14). When the corresponding esters (R^4 = OR) are used, pyrrolin-2-ones are formed (Scheme 6.15).

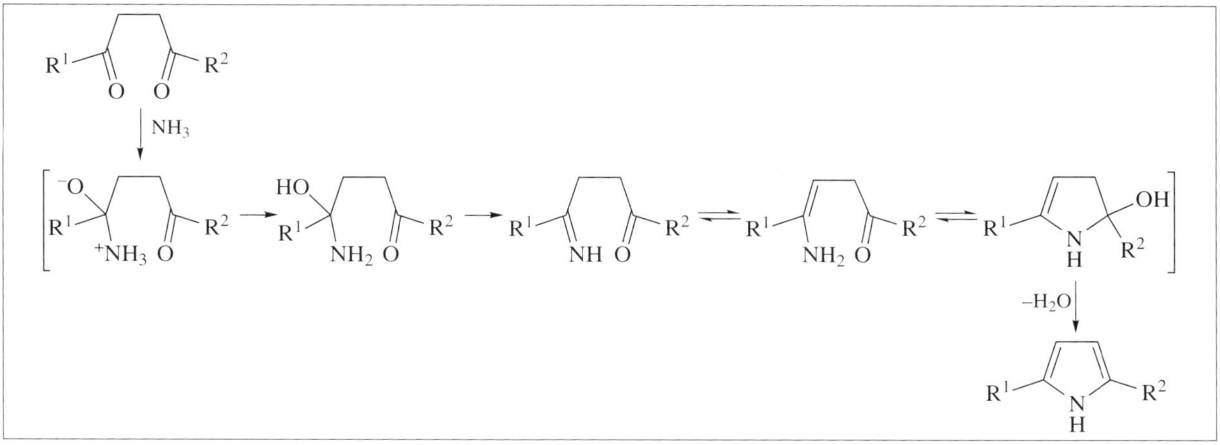

Scheme 6.13

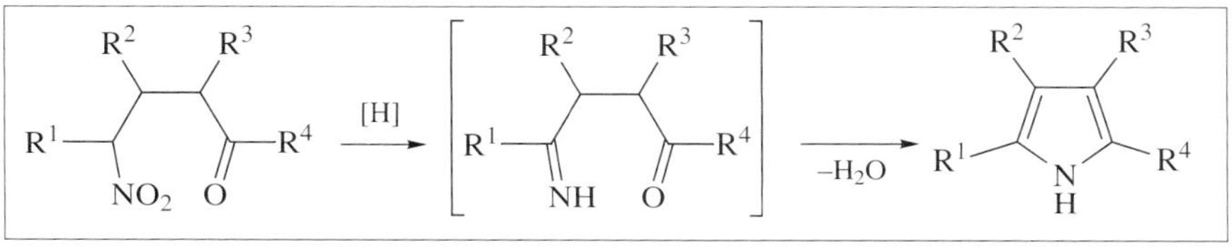

Scheme 6.14

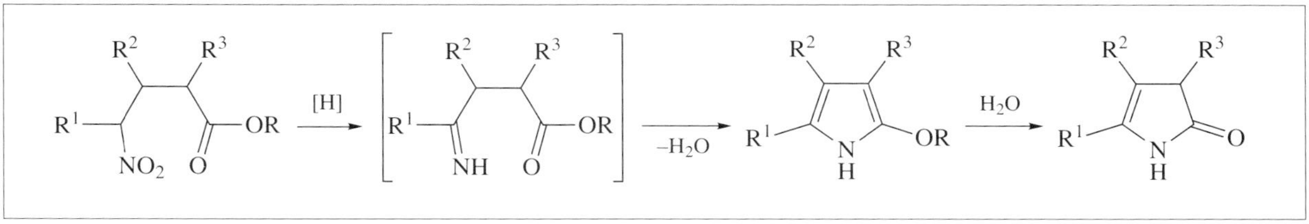

Scheme 6.15

Another preparation involves the reaction of a 2-amino ketone with an activated ester, such as ethyl acetoacetate. The problem here is stopping the amino ketones reacting with themselves to give 1,4-diaza-1,4-dihydrobenzenes (**piperazines**) (Scheme 6.16).

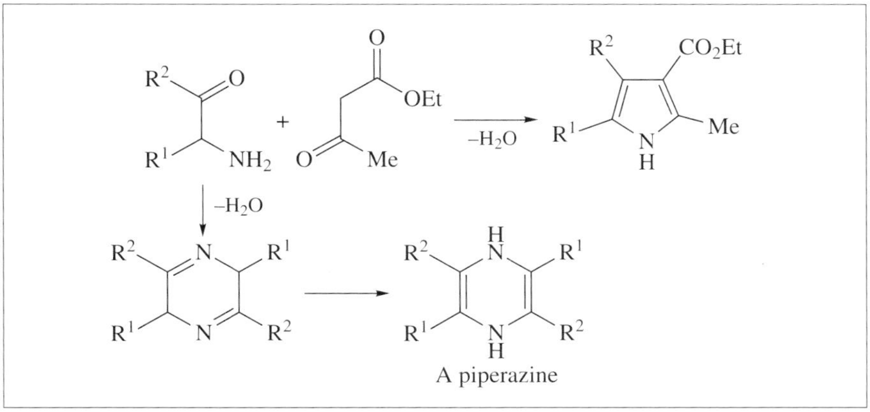

A piperazine

Scheme 6.16

6.1.6 Pyrrolines (Dihydropyrroles) and Pyrrolidine (2,3,4,5-Tetrahydropyrrole)

Dihydropyrroles

There are three dihydro forms of pyrrole: 2,3-dihydro-1*H*-pyrrole (**1-pyrroline**) exists in equilibrium with its imine tautomer (Scheme 6.17). These isomers are more stable than 2,5-dihydro-1*H*-pyrrole (**3-pyrroline**), where no conjugation is possible between the N lone pair electrons and the C=C double bond.

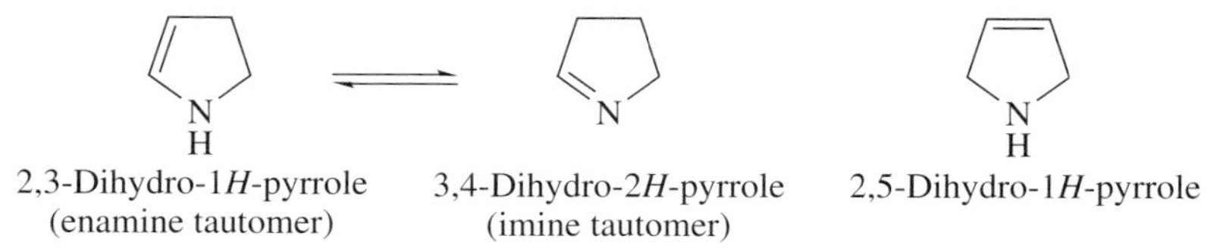

2,3-Dihydro-1*H*-pyrrole   3,4-Dihydro-2*H*-pyrrole   2,5-Dihydro-1*H*-pyrrole
(enamine tautomer)   (imine tautomer)

Scheme 6.17

Although hydrogenation of pyrrole over a rhodium/alumina catalyst gives some 1-pyrroline (Scheme 6.18a), a better method is to dehydrohalogenate *N*-chloropyrrolidine by heating it with alcoholic potassium hydroxide (Scheme 6.18b). 2,5-Dihydro-1*H*-pyrrole, containing 15% pyrrolidine, is obtained by the zinc/hydrochloric acid reduction of pyrrole.

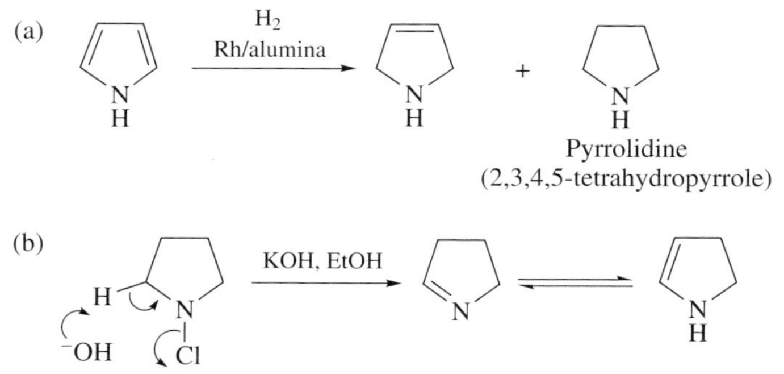

(a) H_2
Rh/alumina

+

Pyrrolidine
(2,3,4,5-tetrahydropyrrole)

(b)   KOH, EtOH

Scheme 6.18

Pyrrolidine

Pyrrolidines can be prepared from 1,4-diaminobutanes by the action of acids (Scheme 6.19).

Pyrrolidine acts as a strong base and it is often used as such in homogenous non-aqueous reactions. Pyrrolidine, like piperidine and other sec-

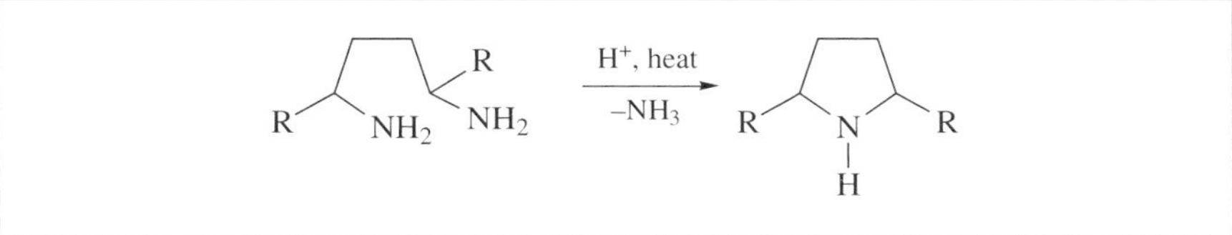

Scheme 6.19

ondary amines, is also used to prepare enamines by reactions with ketones. These enamines are extremely valuable in synthesis, as they allow electrophilic substitution at the α-position of the original ketone, so avoiding potential problems through aldolization reactions and polymerization.

The reaction from an enamine is initiated by the addition of a trace of a strong acid, *e.g.* *p*-toluenesulfonic acid (TsOH, 4-methylbenzene-sulfonic acid), to the ketone and pyrrolidine in a solvent such as toluene. When the mixture is at reflux in a Dean–Stark apparatus, water is liberated and is removed through azeotropic distillation, leaving the enamine in the reaction vessel. After a follow-up reaction between the enamine and a suitable electrophile, an iminium salt is produced that liberates both the α-substituted ketone and pyrrolidine when it is treated with aqueous acid (Scheme 6.20).

The mechanism for the release of the substituted ketone is essentially the reverse of enamine formation.

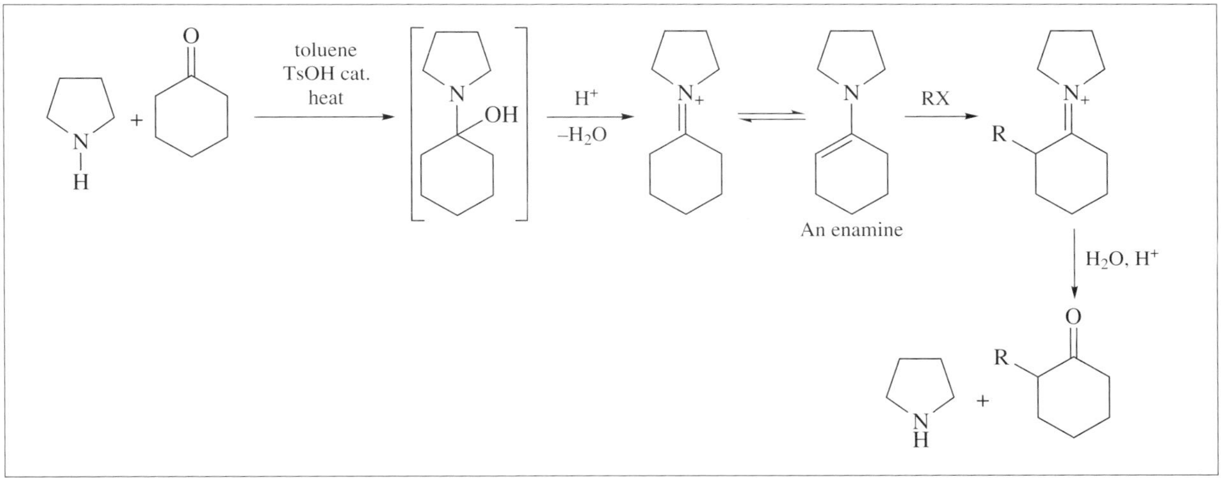

Scheme 6.20

6.2 Furan

6.2.1 Introduction

Furan (oxacyclopentadiene) is also an aromatic compound, although not as aromatic as pyrrole (the oxygen atom is more electron withdrawing).

Like pyrrole, its resonance description requires the delocalization of one of the lone pair electrons of the oxygen atom with the cyclopentadiene unit (Scheme 6.21).

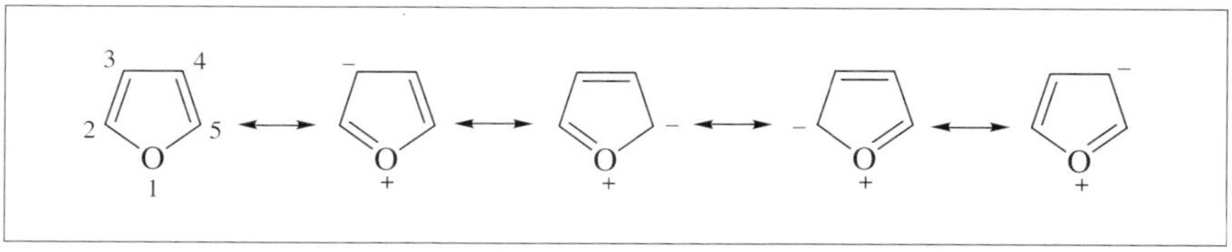

Scheme 6.21

6.2.2 Electrophilic Substitution

For the same reasons as outlined for pyrrole (Section 6.1.2), there is preference for 2- rather than 3-substitution. However, conventional electrophilic reactions, such as nitration, sulfonation, *etc.*, carried out under acidic conditions, are very difficult to control.

In order to achieve 2-nitration, acetyl nitrate may be used as the reagent, but unlike pyrrole a semi-stable adduct, 5-acetoxy-2,5-dihydro-2-nitrofuran, is formed as an intermediate product (Scheme 6.22). This eliminates acetic acid when treated with pyridine. Furan also undergoes initial bromination or chlorination (X = Br or Cl) in ethanol at –40 °C, but then addition of two ethoxyl units occurs with the expulsion of halide ion (Scheme 6.23).

These reactions emphasize the electrophilic character of the σ intermediates formed between furan and positively charged reactants.

Scheme 6.22

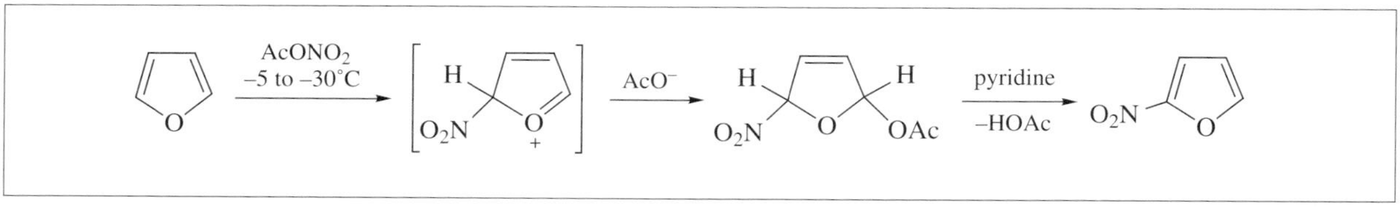

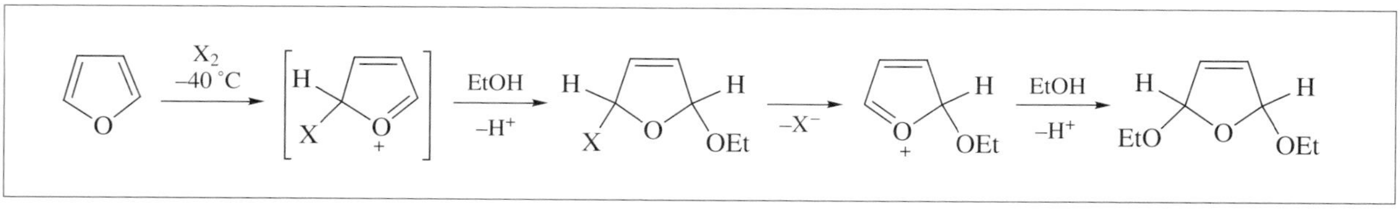

Scheme 6.23   2,5-Disulfonation can be achieved by treating furan with pyridinium-*N*-sulfonate in dichloromethane over several days (Scheme 6.24).

6.2.3 Reactions with Nucleophiles

Furan is resistant to nucleophilic attack, but certain derivatives such as 2-chloromethylfuran undergo side-chain substitution. However, the

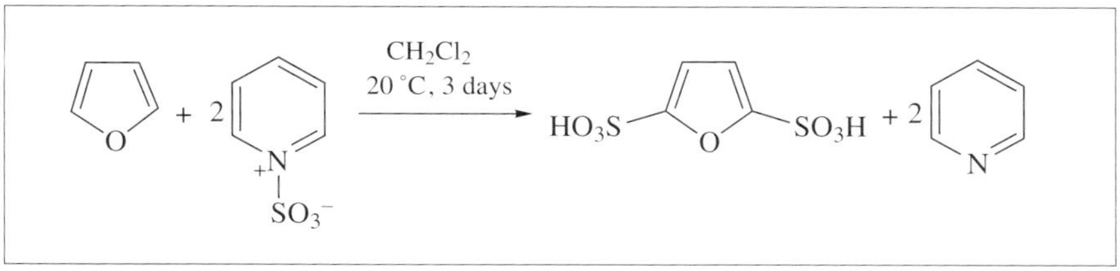

Scheme 6.24

reaction is not a simple one, and treatment with aqueous potassium cyanide gives a mixture of 2-cyanomethylfuran and 2-cyano-5-methylfuran (Scheme 6.25). The first compound might form as the result of a conventional nucleophilic substitution, but equally it could arise (as shown) from the same intermediate as that which gives rise to the second, *i.e.* one formed through elimination of chloride ion and then addition of cyanide ion.

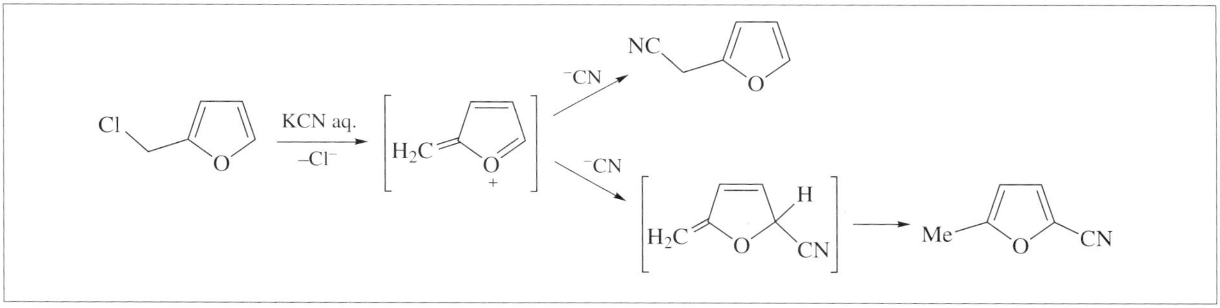

Scheme 6.25

6.2.4 Metallation

Furan reacts with butyllithium in boiling diethyl ether to yield 2-lithiofuran, and further lithiation at C-5 occurs when this product is treated with a second equivalent of the reagent in hexane containing N,N,N',N'-tetramethylethane-1,2-diamine (**TMEDA**) (Scheme 6.26a).

TMEDA is added as a co-solvent to promote the rate of deprotonation of 2-lithiofuran by BuLi.

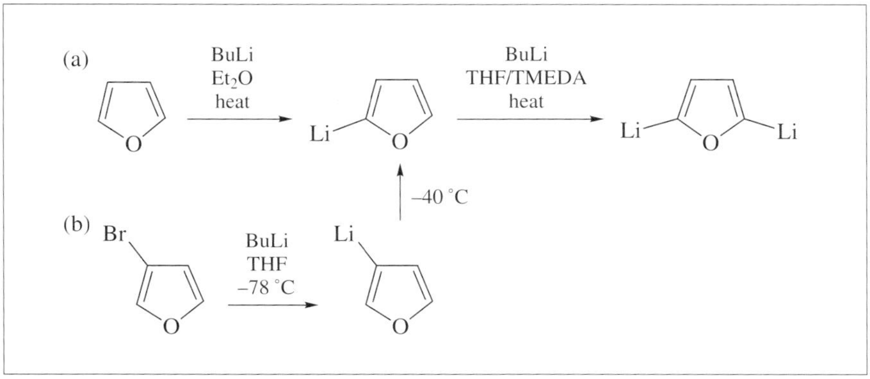

Scheme 6.26

3-Bromofuran undergoes metal–halogen exchange at –78 °C in **THF (tetrahydrofuran)** to yield 3-lithiofuran, but when this product is warmed to about –40 °C it rearranges to 2-lithiofuran (Scheme 6.26b).

6.2.5 Cycloaddition Reactions

The terms **exo** and **endo** are often used to indicate the relative positions of the bridging unit (in this case an oxygen atom) and the residue of the dienophile. When these are on the same face the adduct is referred to as the *exo* form, and when the bridge and the residue are on opposite faces it called the *endo* form. For many pairs of adducts, formed between dienophiles and cyclic dienes, the *exo* product has fewer steric interactions and is the more stable. In some cases, however, secondary electronic effects may overcome steric preferences so that the *endo* (kinetic product) is favoured.

As a result of its reduced aromaticity, relative to pyrrole, furan undergoes [4 + 2] cycloaddition reactions much more readily. It combines as a diene with electron-poor dienophiles to yield **Diels–Alder-type adducts**. Maleic [(Z)-butenedioic acid] anhydride, for example, reacts at room temperature, and the only isolated adduct is the *exo* isomer (the more thermodynamically favoured adduct) (Scheme 6.27a).

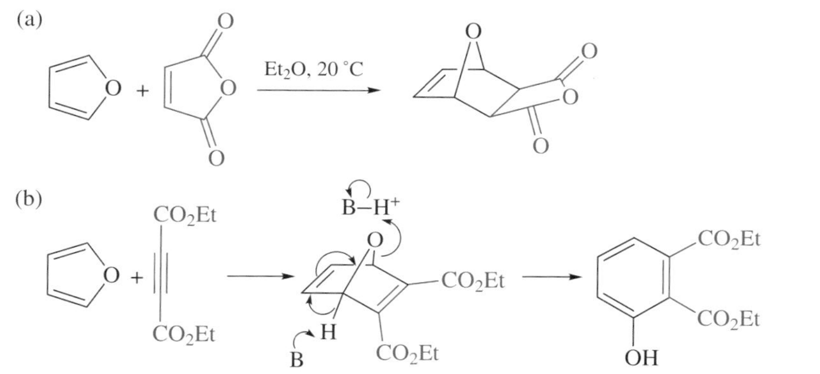

Scheme 6.27

With diethyl acetylenedicarboxylate the adduct formed can be ring opened by base treatment to afford diethyl 3-hydroxyphthalate (3-hydroxybenzene-1,2-dicarboxylate) (Scheme 6.27b).

Carbenes, such as that formed by the photolysis of methyl diazoacetate, may add to one of the carbon–carbon bonds (Scheme 6.28).

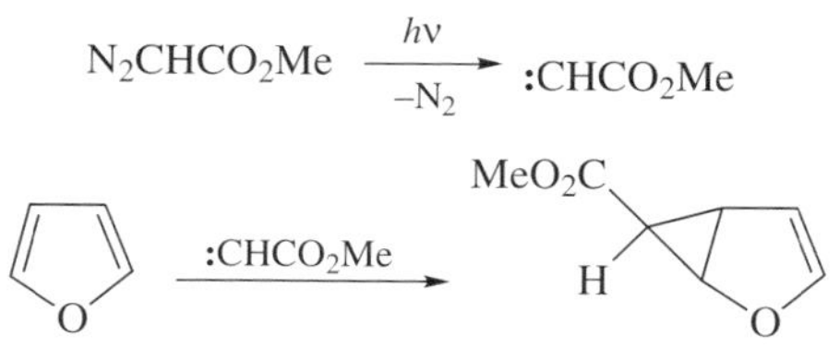

Scheme 6.28

Worked Problem 6.2

Q Suggest a reaction sequence whereby furan reacts with diethyl acetylenedicarboxylate to form a compound that can be partially hydrogenated and heated to give ethene and 3,4-bis(ethoxycarbonyl)furan.

A Furan enters into a cycloaddition reaction with the dienophile diethyl acetylenedicarboxylate to form an adduct. Catalytic reduction occurs preferentially at the more electron-rich double bond of the adduct, and the dihydro product, upon heating, undergoes a reverse Diels–Alder reaction to release ethene and the diester (Scheme 6.29).

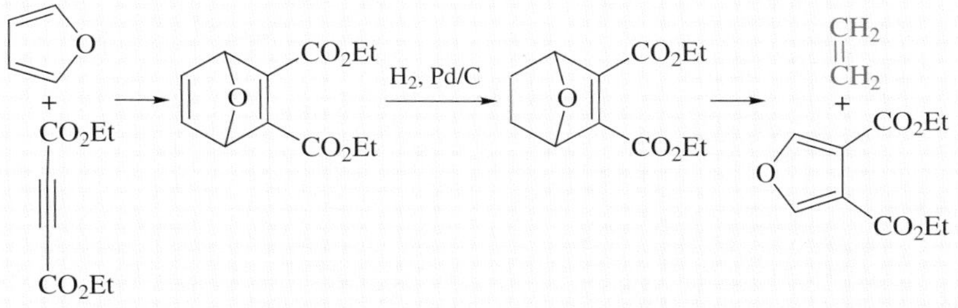

Scheme 6.29

6.2.6 Synthesis

On a commercial scale, furan is obtained from 2-formylfuran (**furfural**, furan-2-carbaldehyde) (see Section 6.2.7) by gas-phase decarbonylation, but in the laboratory, furans can be formed by the cyclodehydration of 1,4-dicarbonyl compounds. Heating in boiling benzene with a trace of p-toluenesulfonic acid as a catalyst in a Dean–Stark apparatus is often effective (Scheme 6.30a).

Scheme 6.30

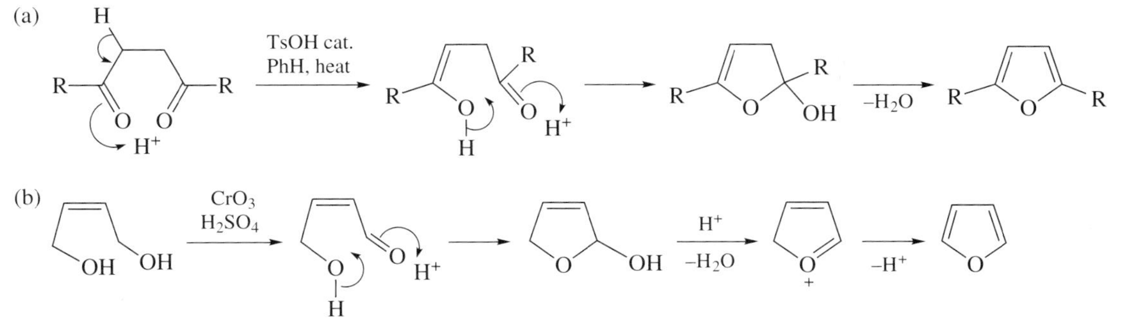

Similarly, 1,4-dihydroxybut-2-enes can oxidized and dehydrated by the action of chromic acid [chromium(IV) oxide in conc. sulfuric acid] (Scheme 6.30b).

6.2.7 Important Derivatives

2-Formylfuran (Furfural, Furfuraldehyde)

This compound is one of the more important furan derivatives, and is commercially available from **pentosans** (polysaccharides) which are present in rice husks, oats and corn residues (furfur is the Latin name for bran!). When treated with sulfuric acid, pentosans decompose into **pentoses**, which then undergo dehydration to the aldehyde (Scheme 6.31).

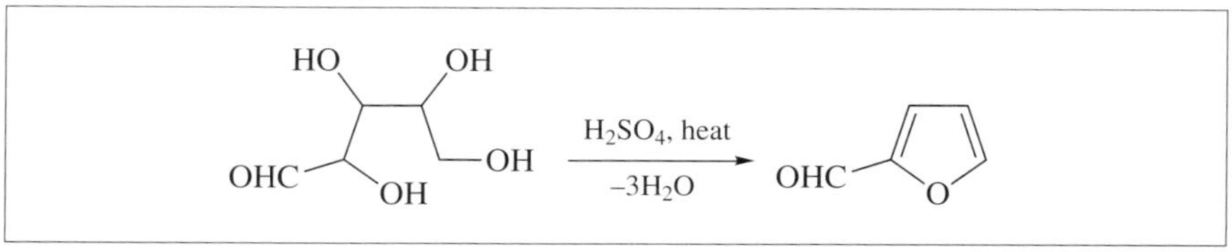

Scheme 6.31

2-Formylfuran behaves in a very similar manner to benzaldehyde and undergoes the usual reactions of an aromatic aldehyde, *e.g.* (i) the **Cannizzaro reaction** with conc. sodium hydroxide to give furan-2-ylmethanol and the sodium salt of furoic acid, (ii) the **Perkin reaction** with acetic anhydride and sodium acetate to yield an aldol product that dehydrates to 3-(furan-2-yl)propenoic acid, and (iii) a condensation with potassium cyanide in alcoholic solution to form **furoin** (under these conditions, benzaldehyde undergoes the benzoin condensation) (Scheme 6.32).

Scheme 6.32

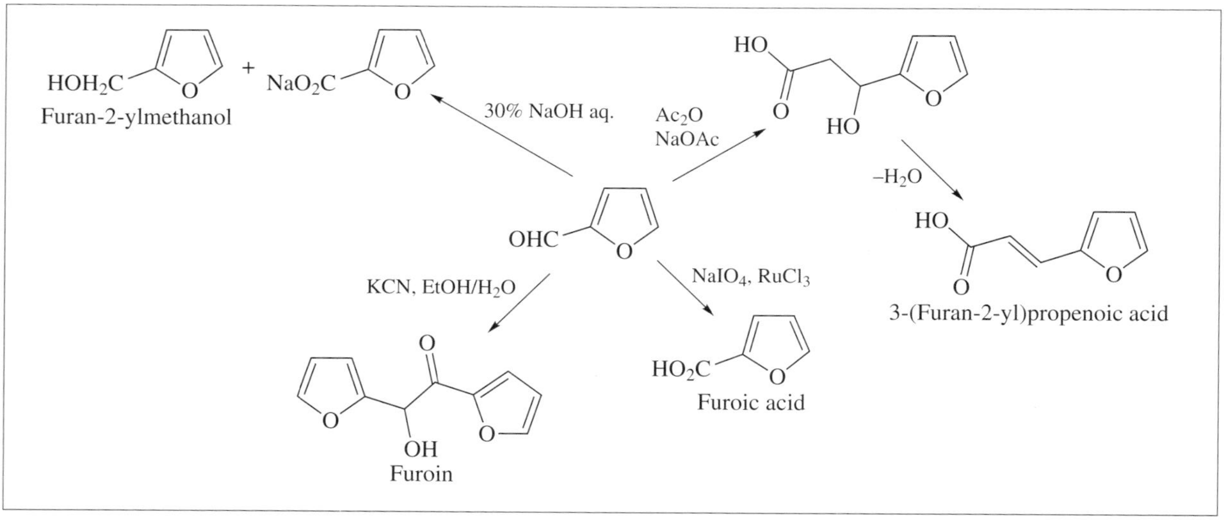

Furoic Acid

2-Formylfuran on oxidation gives **furoic acid** (Scheme 6.32), which shows reactions resembling those of benzoic acid.

6.3 Thiophene

6.3.1 Introduction

Thiophene is present in the benzene fraction from the distillation of coal tar. As with pyrrole and furan, the same type of resonance forms contribute to its overall molecular constitution, and the compound is aromatic in character. There is a difference between thiophene and furan, however, because sulfur is less electronegative than oxygen. Thus, the chemistry of thiophene tends to be closer to that of pyrrole than to that of furan. For example, thiophene does not enter easily into [4 + 2] cycloaddition reactions and quite severe conditions, high pressure (15 bar) and a temperature of 100 °C, are necessary in order to force a cycloaddition between it and maleic anhydride.

Thiophene will form adducts with very strongly electron-depleted dienophiles, such as 1,2-dicyanoethyne and tetrafluorobenzyne.

6.3.2 Electrophilic Substitution

Electrophiles attack thiophene at C-2 and, as for pyrrole, there is a tendency for over-reaction with some traditional electrophilic reagents.

In a commercially useful reaction, sulfuric acid can be used to sulfonate thiophene at room temperature! Benzene is not attacked under these conditions, and needs a reaction with oleum at 60 °C before it is sulfonated. Thiophene-2-sulfonic acid, formed in this way, dissolves in dilute alkali, allowing thiophene in coal tar distillate to be easily removed from benzene. Thiophene can be generated from its sulfonic acid by heating the latter in steam.

Nitration with concentrated nitric acid in acetic anhydride and glacial acetic acid affords a mixture of 2- and 3-nitrothiophenes (in the ratio 6:1) (Scheme 6.33). However, bromination with bromine in diethyl ether and 48% hydrobromic acid, starting at –25 °C and allowing the reaction temperature to rise to –5 °C, gives 2-bromothiophene. If the experiment is initiated at –10 °C and the temperature is then allowed to rise to + 10 °C, 2,5-dibromothiophene is formed. 2,3,5-Tribromination occurs in 48% hydrobromic acid, starting the experiment at room temperature and allowing the temperature to rise to 75 °C. Acetylation with acetyl chloride in the presence of the Lewis acid tin(IV) chloride gives 2-acetylthiophene, and efficient 2-formylation takes place under Vilsmeier conditions (see Section 6.1.2).

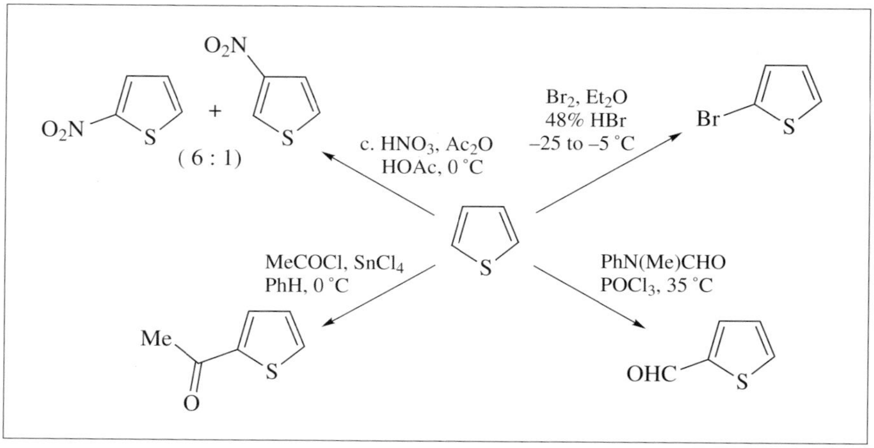

Scheme 6.33

6.3.3 Oxidation

Oxidation of thiophene with *m*-chloroperbenzoic acid (MCPBA, 3-chloroperoxobenzoic acid) probably gives first unstable **thiophene 1-oxide**, and then **thiophene 1,1-dioxide** (Scheme 6.34a). 2,5-Diarylthiophenes can be oxidized to the corresponding 1-oxides with 30% aqueous hydrogen peroxide in trifluoroacetic acid and dichloromethane.

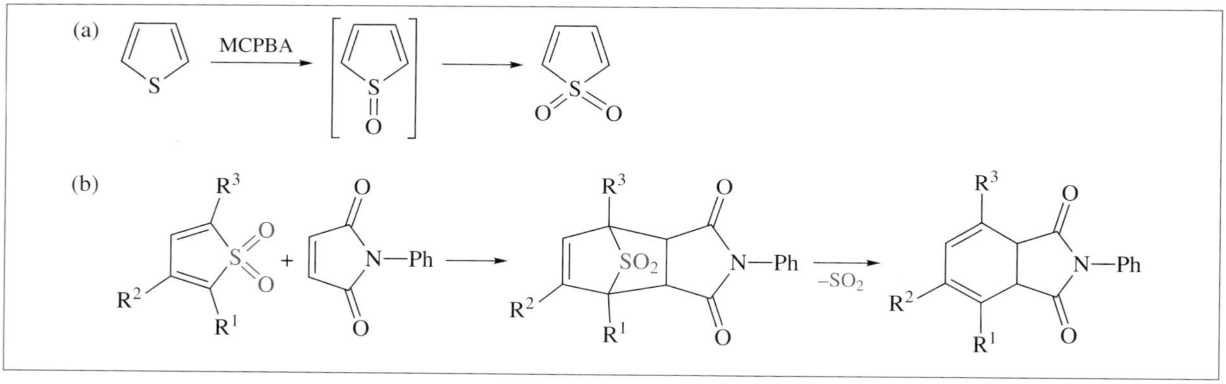

Scheme 6.34

Thiophene 1,1-dioxides, unlike most other thiophene derivatives, are dienes and combine with dienophiles to form adducts that are prone to retro-cycloaddition, extruding sulfur dioxide in the process (Scheme 6.34b).

Worked Problem 6.3

Q Why should thiophene 1,1-dioxide behaves as a diene in cycloaddition reactions, whereas benzo[*b*]thiophene 1,1-dioxide acts as a dienophile?

A The strongly electronegative sulfone unit in thiophene 1,1-dioxide destroys the aromaticity of the thiophene component; hence the compound behaves as a diene. Even though thiophene 1,1-dioxide is as an electron-poor diene, it is very reactive and cycloadditions do take place between it and electron-poor dienophiles. For example, 3,4-di-*t*-butylthiophene 1,1-dioxide forms an adduct when it is heated with phenyl vinyl (ethenyl) sulfoxide (Scheme 6.35a). As often happens with adducts of this type, sulfur dioxide is expelled and 1,2-di-*t*-butylbenzene is eventually formed (for a review of the detailed chemistry of thiophene 1,1-dioxides, see Gronowitz[1]). A similar reduction in the aromaticity of the heterocycle affects benzo[*b*]thiophene 1,1-dioxide, but here the benzene resonance is maintained and the ring fusion blocks a double bond of the thiophene unit. As a result, the compound behaves as an electron-poor dienophile, forming an adduct with, for example, cyclopentadiene (Scheme 6.35b).

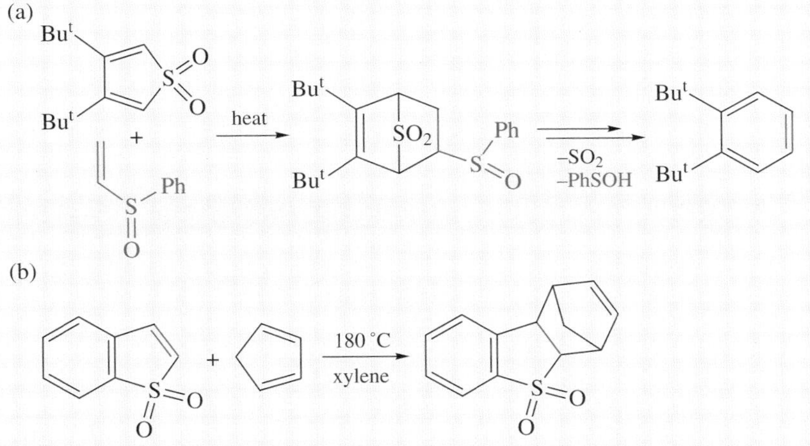

Scheme 6.35

6.3.4 Metallation

Butyllithium effects C-2 lithiation, and the lithium derivative can then be reacted with electrophiles (this is a good way to synthesize a wide variety of 2-substituted thiophenes) (Scheme 6.36).

Both 2-bromo- and 3-bromothiophenes and their derivatives undergo metal–halogen exchange with butyllithium and form Grignard reagents when reacted with magnesium turnings suspended in diethyl ether or THF.

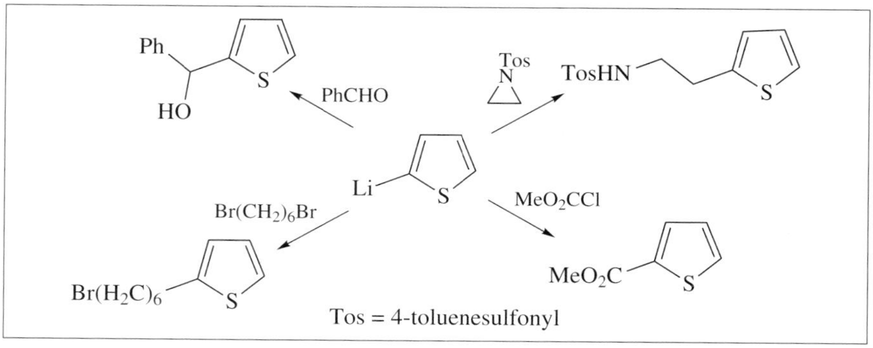

Scheme 6.36

6.3.5 Synthesis

This is often achieved from 1,4-dicarbonyl compounds in a procedure similar to that used to form furans but using phosphorus pentasulfide, or **Lawesson's reagent** (see below), to cause a transposition from carbonyl to thiocarbonyl groups, prior to cyclization and loss of hydrogen sulfide (Scheme 6.37).

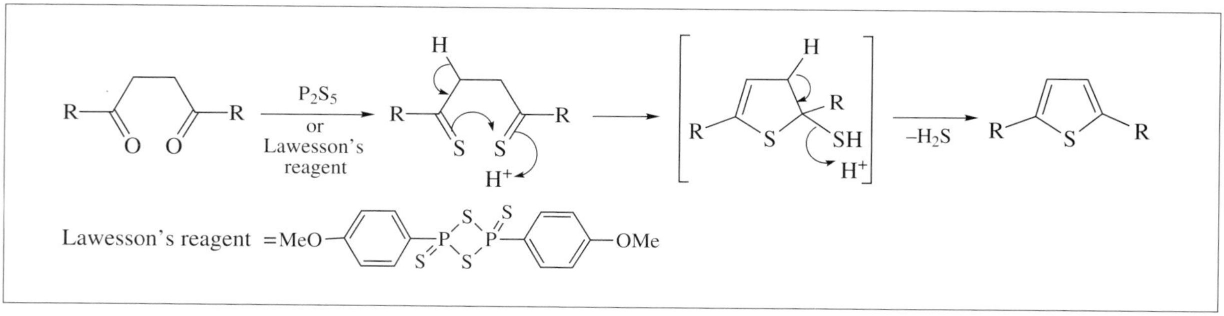

Scheme 6.37

Summary of Key Points

1. Pyrrole, furan and thiophene are all aromatic compounds, furan being the least aromatic. All react with electrophiles at C-2(5).

2. All react with electrophiles at C-2(5).

3. Pyrrole can be deprotonated by strong bases, forming the resonance-stabilized pyrrolyl anion. The anion is susceptible to reactions with electrophiles, the location depending upon the reagent and the conditions.

4. Furan can act as a diene in cycloaddition reactions, but pyrrole only does so when activated by *N*-substitution with an electron-withdrawing group. Thiophene 1,1-dioxide is also an effective diene component in [4 + 2] cycloadditions.

Problems

1. Why should 1-[tris(propan-2-yl)silyl]pyrrole react with *N*-bromosuccinimide (a source of Br_2) to give 3-bromo-1-[tris(propan-2-yl)silyl]pyrrole, rather than the 2-isomer?

2. Suggest mechanisms for reactions whereby pyrrole is ring opened to the dioxime of butane-1,4-dial by the action of hydroxylamine in ethanol and hydrochloric acid.

3. Suggest syntheses of compounds (a)–(c) from suitable starting materials.

4. 2-Formylfuran reacts with potassium cyanide in alcohol solution to give a product containing two furan nuclei. What is this product and how does it form? Can you suggest an alternative synthesis of the compound from 2-formylfuran using propane-1,3-dithiol as one reagent? (*Hint*: consider an acyl anion equivalent approach).

Reference

1. S. Gronowitz, *Phosphorus, Sulfur, Silicon*, 1993, **74**, 113.

Further Reading

Pyrroles
R. A. Jones and G. P. Bean, *The Chemistry of Pyrrole*, Academic Press, London, 1977.

M. P. Sammes and A. R. Katritzky, *The 2H- and 3H-Pyrroles*, in *Adv. Heterocycl. Chem.*, 1982, **32**, 233.

R. A. Jones (ed.), *Pyrroles*, in *The Chemistry of Heterocyclic Compounds*, ed. A. Weissberger and E. C. Taylor, vol. 48, parts 1 and 2, Wiley, New York, 1990–1992.

A. V. Patel and T. A. Crabb, *Pyrroles, Pyrrolines and Pyrrolidines*, in *Rodd's Chemistry of Carbon Compounds*, 2nd. edn., 2nd suppl., vol. IVA, chap. 4, ed. M. Sainsbury, Elsevier, Amsterdam, 1997.

D. StC. Black, *1H-Pyrroles*, in *Science of Synthesis*, vol. 9, ed. G. Maas, Thieme, Stuttgart, 2001, chap. 9.13.

Furans

A. P. Dunlop and F. N. Peters, *The Furans*, Reinhold, New York, 1952.

P. Bosshar and E. H. Eugster, *The Development of the Chemistry of Furans, 1952–1963*, in *Adv. Heterocycl. Chem.*, 1966, **7**, 377.

F. M. Dean, *Recent Advances in Furan Chemistry*, Parts 1 and 2, in *Adv. Heterocycl. Chem.*, 1982, **30**, 167; 1983, **31**, 237.

D. T. Hurst, *Furans, Benzofurans, Isofurans and their Reduced Forms*, in *Rodd's Chemistry of Carbon Compounds*, 2nd. edn., 2nd suppl., vol. IVA, chap. 2, ed. M. Sainsbury, Elsevier, Amsterdam, 1997.

B. König, *Furans*, in *Science of Synthesis*, vol, 9, ed. G. Maas, Thieme, Stuttgart, 2001, chap. 9.9.

Thiophenes

D. E. Wolf and K. Folkers, *The Preparation of Thiophenes and Tetrahydrothiophenes*, in *Org. React.*, 1951, **6**, 410.

S. Gronowitz (ed.), *Thiophene and its Derivatives*, in *The Chemistry of Heterocyclic Compounds*, ed. A. Weissberger and E. C. Taylor, vol. 44, Wiley, New York, 1985.

B. Iddon, *Cycloaddition, Ring Opening and Other Novel Reactions of Thiophenes*, in *Heterocycles*, 1983, **20**, 1127.

K. Hale and S. Manaviazar, *Thiophenes, Hydrothiophenes, Benzothiophenes, and Related Compounds*, in *Rodd's Chemistry of Carbon Compounds*, 2nd edn., 2nd suppl., vol. IVA, chap. 3, ed. M. Sainsbury, Elsevier, Amsterdam, 1997.

J. Schatz, *Thiophenes, Thiophene 1,1-Dioxides, and Thiophene 1-Oxides*, in *Science of Synthesis*, vol, 9, ed. G. Maas, Thieme, Stuttgart, 2001, chap. 9.10.

7

Benzo[*b*]pyrrole, Benzo[*b*]furan and Benzo[*b*]thiophene

Aims

By the end of this chapter you should understand:

- The effect that the fusion of a benzene ring has upon the reactions of the five-membered parent heterocycle
- The general reactions of benzo[*b*]pyrrole, -furan and -thiophene
- The main methods of synthesis and the mechanisms of the reactions involved, particularly those for benzo[*b*]pyrrole
- The reactivity of certain indole derivatives, such as oxindole and indoxyl

7.1   Indole (Benzo[*b*]pyrrole)

7.1.1 Introduction

Indole is the parent of a very large number of alkaloids and medicinally important compounds. Its reactions, and particularly the synthesis of complex derivatives, occupy central stage in heterocyclic chemistry.

The definition of aromaticity conceived by Hückel strictly applies to monocyclic ring systems, but indole, constructed from the fusion of benzene and pyrrole, behaves as an aromatic compound, like quinoline and isoquinoline. The ring fusion, however, affects the properties of both components. This is reflected in the valence bond description of indole, shown in Scheme 7.1, where one canonical representation shows electron density shared between N-1 and C-3 in the pyrrole unit (implying enamine character). Note that although other canonical forms can be drawn, where the lone-pair electrons are delocalized into the benzenoid ring, their energy content is relatively high and they are of limited importance.

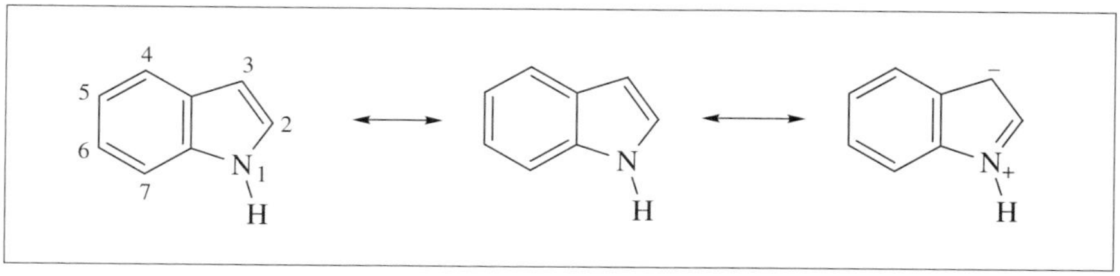

Scheme 7.1

7.1.2 Electrophilic Substitution

C-3 *versus* C-2

Electrophiles attack indole at C-3, rather than at C-2 (Scheme 7.2). This is the opposite result to that observed for pyrroles, but can be explained if the intermediates for each type of reaction are considered. For a reaction at C-3, the energy of activation of the intermediate is lowered because it is possible to delocalize the positive charge through resonance involving the nitrogen lone pair of electrons. This favourable situation is not possible in the corresponding intermediate for attack at C-2. Any attempt to delocalize the positive charge would now disrupt the 6π-electron system of the benzene ring.

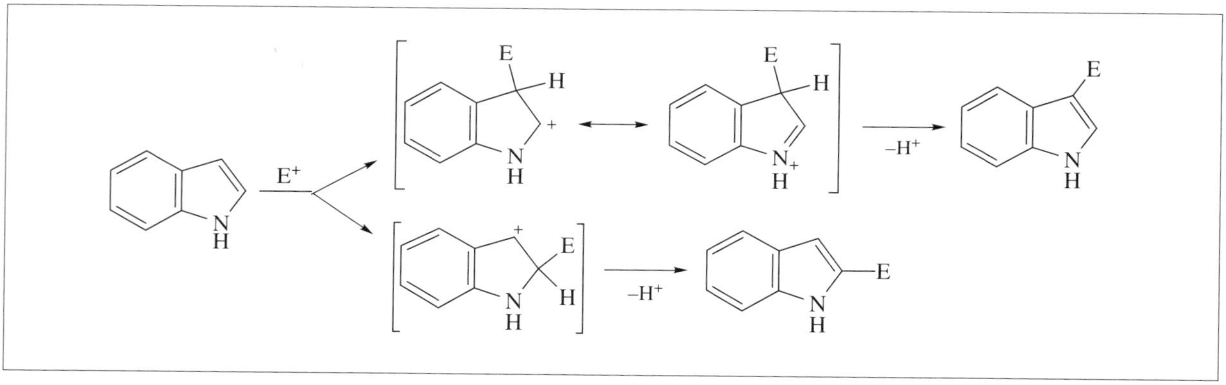

Scheme 7.2

2,3-Disubstituted Indoles

In times past it was thought that indoles already bearing an alkyl substituent at C-3 were further alkylated by direct attack at C-2. However, although 2,3-dialkylindoles are readily formed the reaction *still involves attack at C-3*. This can be demonstrated by the example in Scheme 7.3, where 3-(4′-hydroxybutyl)indole, containing an isotopic label located at C-1′, is treated with boron trifluoride in diethyl ether. *Two* 1,2,3,4-tetrahydrocarbazoles (1,2,3,4-tetrahydrodibenzo[b,d]pyrroles) are formed in a ratio of 1:1. These differ only in the position of the label. This result indicates that a **3,3-spiroindoleninium** intermediate is formed first, and this then undergoes rearrangement of either bond a or bond b to C-2. As the two bonds a and b are identical, equal amounts of the tetrahydrocarbazoles are formed. This is a general type of reaction for indoles, but when two different groups are present at C-3 in the intermediate their ability to migrate may not be the same. Hence, after rearrangement the proportions of products may no longer be 1:1.

A **spiro** compound contains a carbon atom that is common to two ring systems.

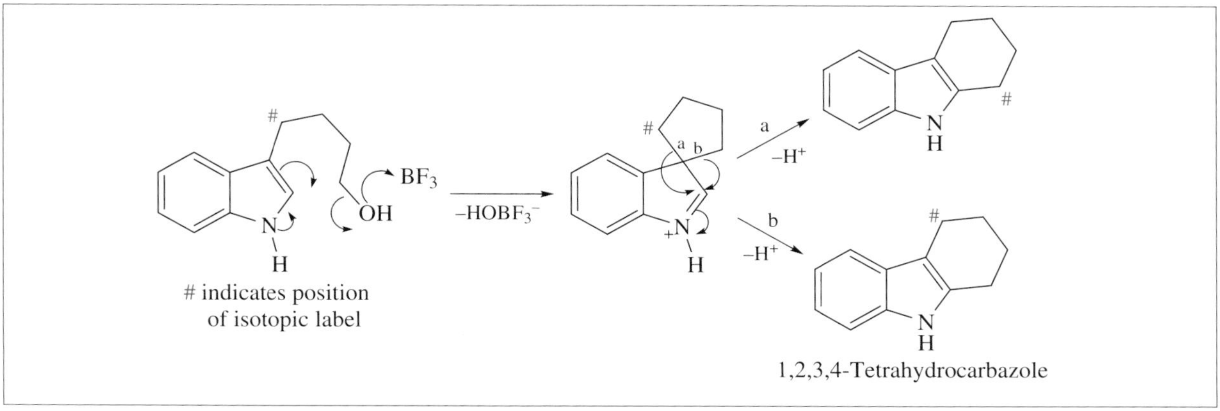

# indicates position of isotopic label

1,2,3,4-Tetrahydrocarbazole

Protonation of indole has been shown to give the indoleninium (3*H*-indolium) cation, and this is presumably formed under the conditions commonly used for electrophilic substitution reactions in the benzene series. For indole these conditions are often unsatisfactory and, for example, nitration of indole with nitric acid gives polymers. On the other hand, *N*-alkylindoles can be nitrated with concentrated nitric acid and acetic anhydride at –70 °C to afford 1-alkyl-3-nitroindoles.

Scheme 7.3

The term **indolenine** (or 3*H*-indole) is in common use; the corresponding cation is named **indoleninium** (or 3*H*-indolium). *Note*: 2,3-dihydroindole is called **indoline**:

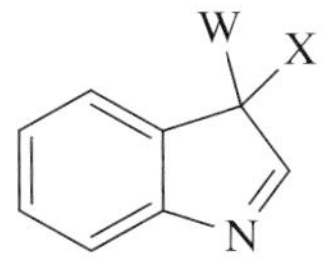

Indolenine (3*H*-indole)

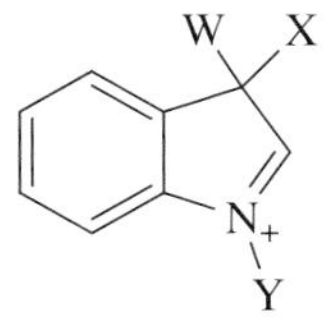

Indoleninium (3*H*-indolium)

3-Substituted Indoles

Sulfonation of indole with pyridinium-*N*-sulfonate yields indolyl-3-sulfonic acid, and bromine in pyridine at 0 °C affords 3-bromoindole (Scheme 7.4). Acetylation with a heated mixture of acetic anhydride and acetic acid gives 1,3-diacetylindole. Methylation requires heating with methyl iodide in DMF (*N*,*N*-dimethylformamide) at 80–90 °C and yields 3-methylindole. This compound reacts further, giving 2,3-dimethylindole and finally 1,2,3,3-tetramethyl-3*H*-indoleninium iodide.

Vilsmeier Formylation

When DMF and phosphorus oxychloride are reacted together in the Vilsmeier reaction, the *N*,*N*-dimethylamino(chloro)methyleniminium cation is generated, and this reacts with indole at 5 °C to give 3-(*N*,*N*-dimethylaminomethylene)indolenine (see Section 6.1.2). When hydrolysed by treatment with dilute sodium hydroxide, this gives an excellent yield of 3-formylindole (Scheme 7.5).

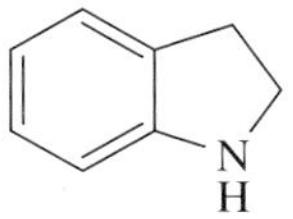

Indoline (2,3-dihydroindole)

Scheme 7.4

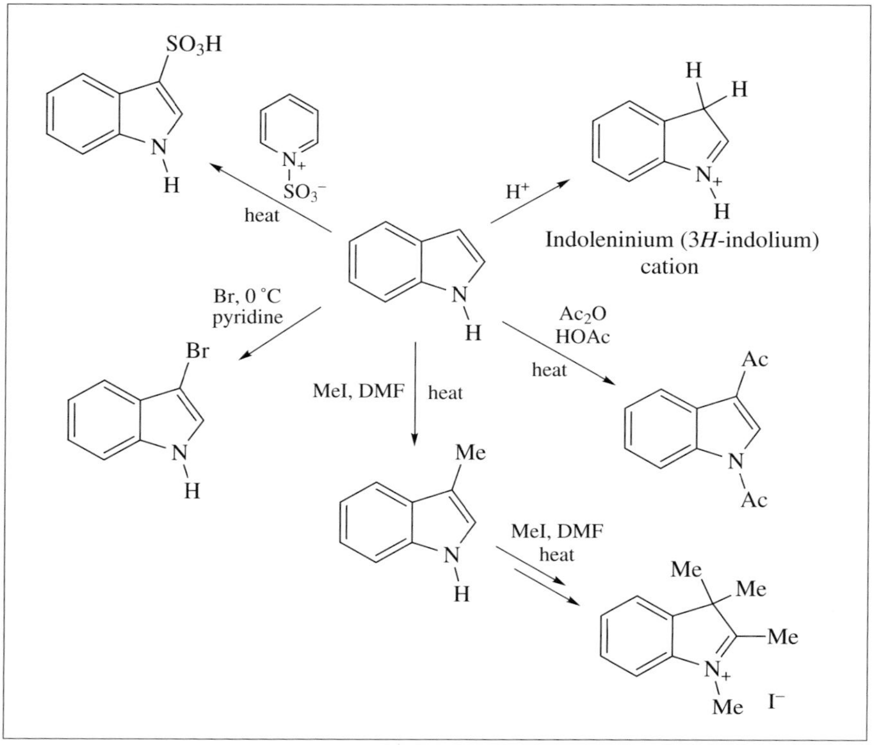
SO3H
heat
H+
Indoleninium (3H-indolium) cation
Br, 0 °C pyridine
Br
MeI, DMF
heat
Ac2O
HOAc
heat
Ac
Me
MeI, DMF
heat
I−

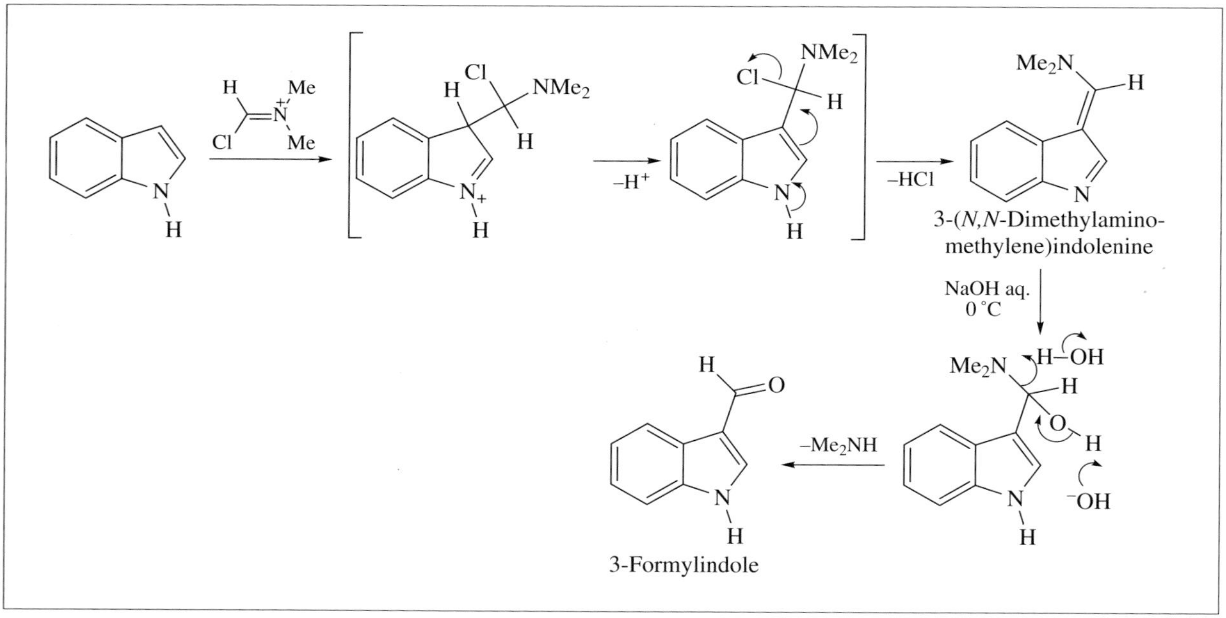
−H+
−HCl
3-(N,N-Dimethylamino-methylene)indolenine
NaOH aq.
0 °C
−Me2NH
3-Formylindole

Scheme 7.5

Reaction with Triethyl Orthoformate (Triethoxymethane)

Triethyl orthoformate is often used in reactions with enolates and carbanions to form diethyl acetals that on treatment with dilute acid give the corresponding formyl derivatives. However, when indole is heated at 160 °C with triethyl orthoformate the locus of reaction is at N-1 rather than at C-3, and 1-(diethoxymethyl)indole is formed (Scheme 7.6). The *N*-substituent is easily removed by acidic hydrolysis to reform indole.

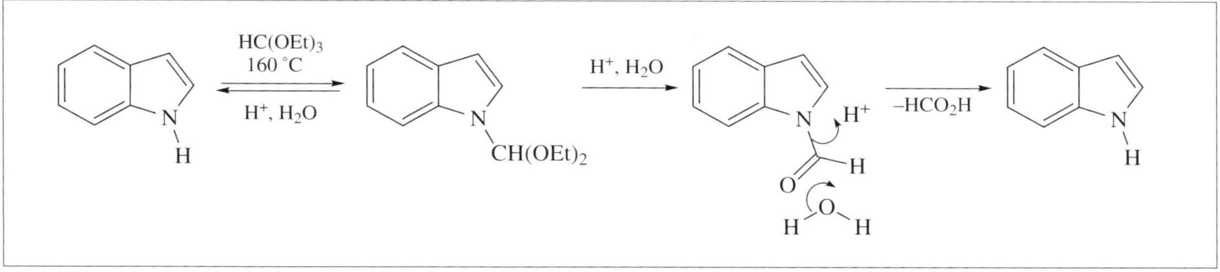

Mannich Reaction

Scheme 7.6

The mechanism of the **Mannich reaction** is similar to that of the Vilsmeier reaction as the electrophile is also a methyleniminium cation, formed this time from a condensation of dimethylamine and formaldehyde in acetic acid solution (Scheme 7.7a). This reacts with indole to yield 3-(*N*,*N*-dimethylaminomethyl)indole (although not shown, it is possible that initial attack occurs at N-1 and rearrangement of the side chain to C-3 takes place in a follow-up step) (Scheme 7.7b).

Scheme 7.7

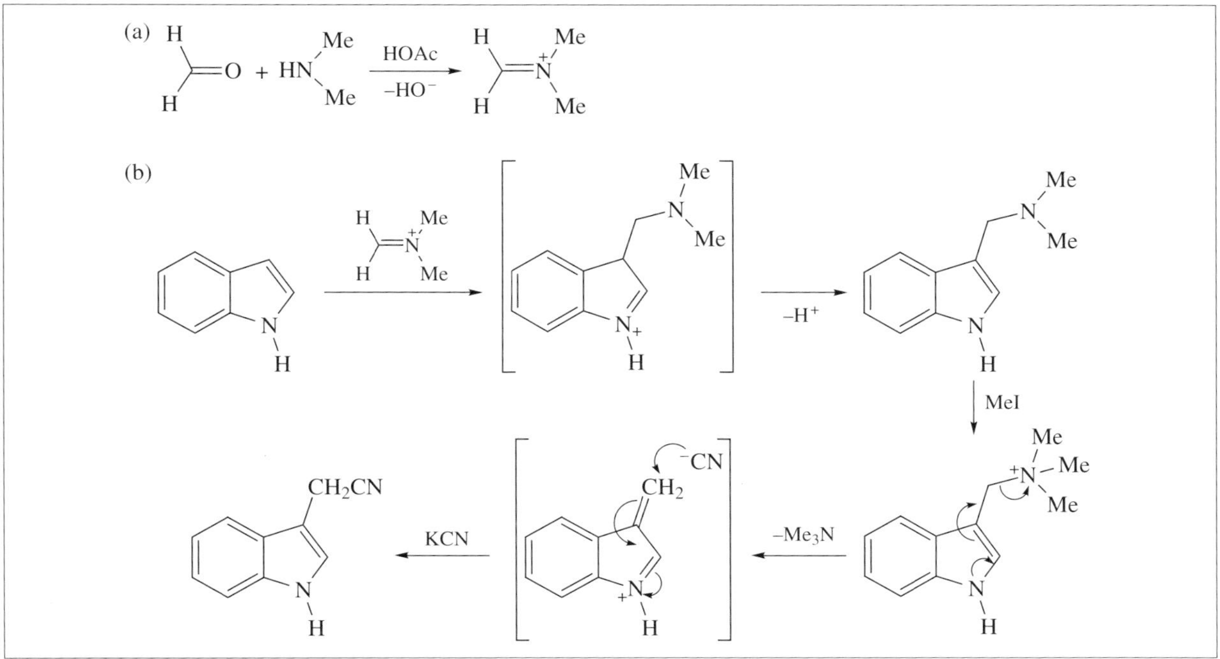

To make use of the Mannich reaction it is possible to methylate the N-atom of the new side chain and eliminate trimethylamine. The product, a 3-methyleneindoleninium salt, can then be trapped with suitable nucleophiles. In the example shown in Scheme 7.7b, cyanide ion is used, and reduction of the resultant nitrile yields the important amine **tryptamine**. Indol-3-ylacetonitrile is also the source of indol-3-ylacetic acid and other biologically useful compounds (see Section 7.1.7).

Worked Problem 7.1

Q What is the probable reaction mechanism for the reaction of indole and hexane-2,4-dione that yields 1,4-dimethylbenzo[*b*]indole (1,4-dimethylcarbazole)?

A The reaction is likely to proceed in a step-wise fashion, forming first a 3-hydroxyalkylindole, which then reacts again at C-3 to give a spiro intermediate (Scheme 7.8). Migration of one of the substituents (they are, of course, identical so there is no preference) then gives a 1,4-dihydroxy-1,2,3,4-tetrahydrocarbazole that dehydrates and aromatizes into the final product.

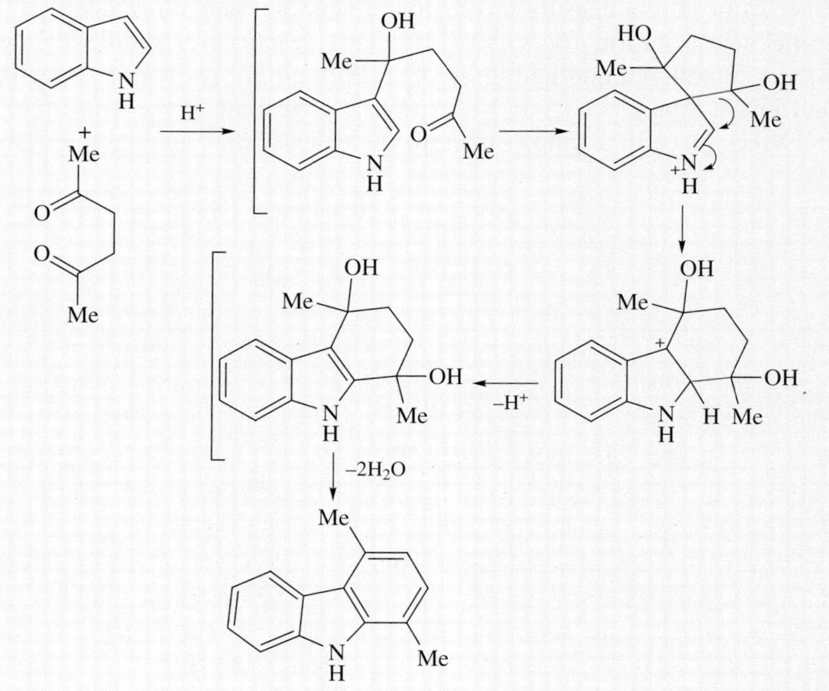

Scheme 7.8

7.1.3 Formation of the Indolyl Anion; N-1 *versus* C-3 Substitution

The indole anion is easily formed by reactions with bases such as sodium hydride, sodamide, Grignard reagents or alkyllithiums. Although the indolyl anion is resonance stabilized, the nature of the cation has an influence upon future reactions (as does the solvent used). Thus, if the conjugate cation is not easily polarized, *e.g.* a sodium ion (or potassium ion), the indolyl anion is attacked at the site of highest electron density, *i.e.* at N-1. However, if the metal in the cation is magnesium, then it is assumed that partial covalent bonding to nitrogen prevents attack there. Now the electrophilic attack is diverted to C-3.

N-Alkylation, -acylation and -sulfonation are also promoted by a polar solvent, such as HMPA (hexamethylphosphoric triamide). This acts to solvate the ions (promoting dissociation), but in a non-polar solvent like diethyl ether or tetrahydrofuran (THF), attack by most carbon electrophiles upon indolylmagnesium bromide proceeds at C-3 (Scheme 7.9).

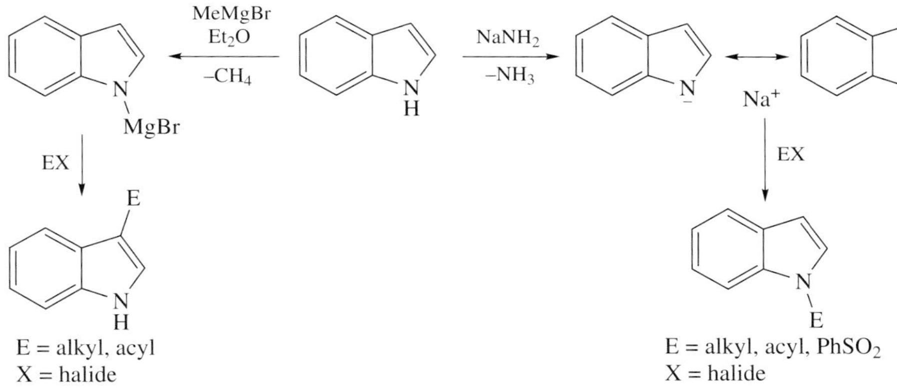

C-2 Lithiation

In order to access the C-2 position, indirect methods of reaction are used, and a common procedure is to *N*-sulfonate indole with sodium hydride and benzenesulfonyl chloride and then to treat the derived sulfonate with butyllithium. C-2 deprotonation and lithiation occur (facilitated by chelation to the sulfonyl group) and the intermediate, without isolation, can then be reacted with a wide range of electrophiles at this site. Finally, the sulfonyl group can be hydrolysed off in a separate step to form the desired product (Scheme 7.10).

Scheme 7.9

The word **chelation** derives from the Greek word *chela*, meaning claw-like. Here, lithium is held between C-2 and the SO$_2$Ph group by partial ionic bonds.

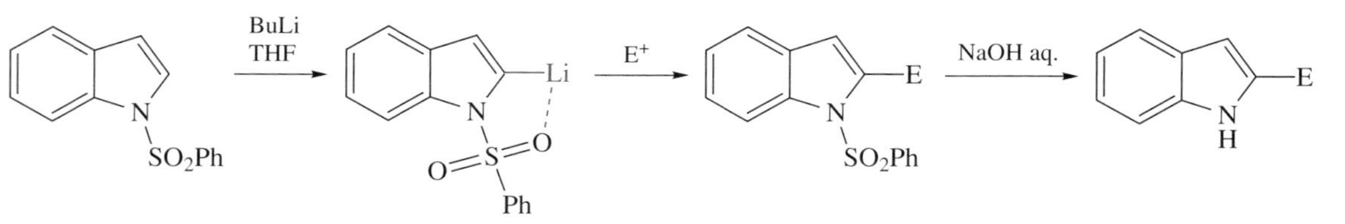

Scheme 7.10

Worked Problem 7.2

Q Outline a synthesis of 2-acetyl-1-(phenylsulfonyl)indole starting from indole.

A Indole, when treated with one equivalent of sodamide and then with benzenesulfonyl chloride, gives 1-(phenylsulfonyl)indole. The *N*-sulfonyl substituent activates the H-2 to deprotonation by butyllithium and stabilizes the lithium derivative by chelation. This '*ortho* lithiation' process facilitates subsequent acetylation at this site by acetyl chloride, affording 2-acetyl-1-(phenylsulfonyl)indole (Scheme 7.11).

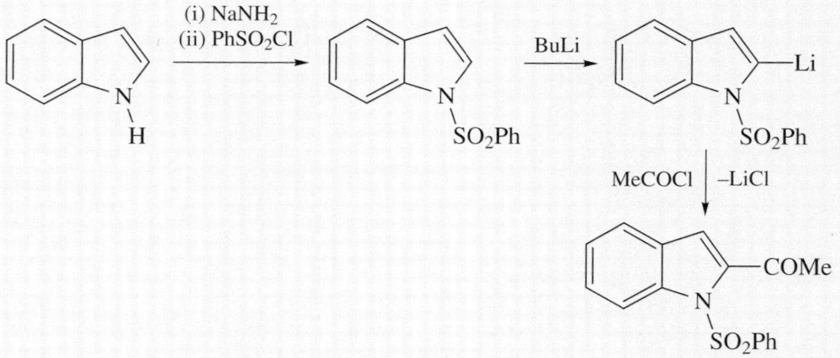

Scheme 7.11

7.1.4 Ring Expansion with Dichlorocarbene

Indole can be reacted with dichlorocarbene to yield 3-chloroquinoline (Scheme 7.12). Initially, the carbene adds across the C-2–C-3 double bond to form a cyclopropanoindole; this product then ring expands with elimination of hydrogen chloride (*cf.* pyrrole, Section 6.1.3).

7.1.5 Reduction

Indole can be reduced under Birch conditions (lithium in liquid ammonia containing a hydrogen donor, *e.g.* methanol) to give a 4:1 mixture

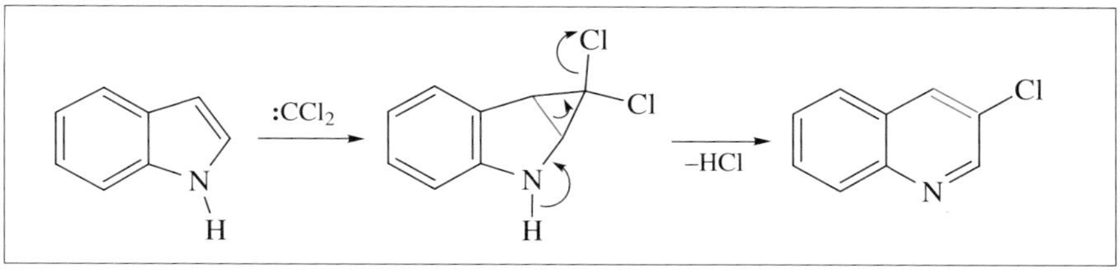

Scheme 7.12

of 4,7-di- and 4,5,6,7-tetrahydroindoles (Scheme 7.13a). Sodium cyanoborohydride (NaBH$_3$CN) in acetic acid, however, forms indoline (2,3-dihydroindole). In this reduction, 3-protonation gives the indoleninium salt, which is then 'set up' to undergo hydride transfer at C-2 from the boron reagent (Scheme 7.13b).

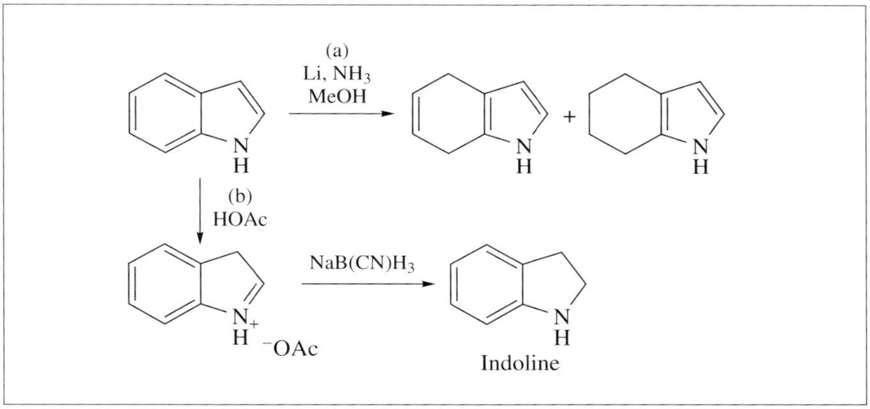

Scheme 7.13

7.1.6 Synthesis

Fischer Indolization Reaction

Most indoles are synthesized by the Fischer indolization reaction. Here a phenylhydrazine is first reacted with an aldehyde, or ketone, carrying an α-methylene group (not acetaldehyde). The corresponding hydrazone is then treated with an acid, often hydrochloric acid. Ring closure occurs, through a [3,3]-sigmatropic change, and ammonia (as the ammonium cation) is lost (Scheme 7.14).

A variant on the traditional method is to use a phenylhydroxylamine instead of a phenylhydrazine. For example, *N*-(benzyloxycarbonyl)-phenylhydroxylamine can be reacted with methyl propynoate to give an adduct that spontaneously undergoes a [3,3]-sigmatropic shift and eventually yields 1-(benzyloxycarbonyl)-3-(methoxycarbonyl)indole (Scheme 7.15).

The mechanism of the reaction has been much studied, but even now the precise details are uncertain and the extent of protonation of the reactant and intermediates depends upon the conditions used (in some cases, aprotic conditions have been used). Labelling studies (with ^{15}N) show that the NH$_2$ unit of the phenylhydrazine is eventually lost as ammonia. Note: the term **sigmatropic** indicates that a single (σ) bond is moving to a new location. In this case, both termini of the new bond are three atoms away from their original positions: a [3,3]-shift.

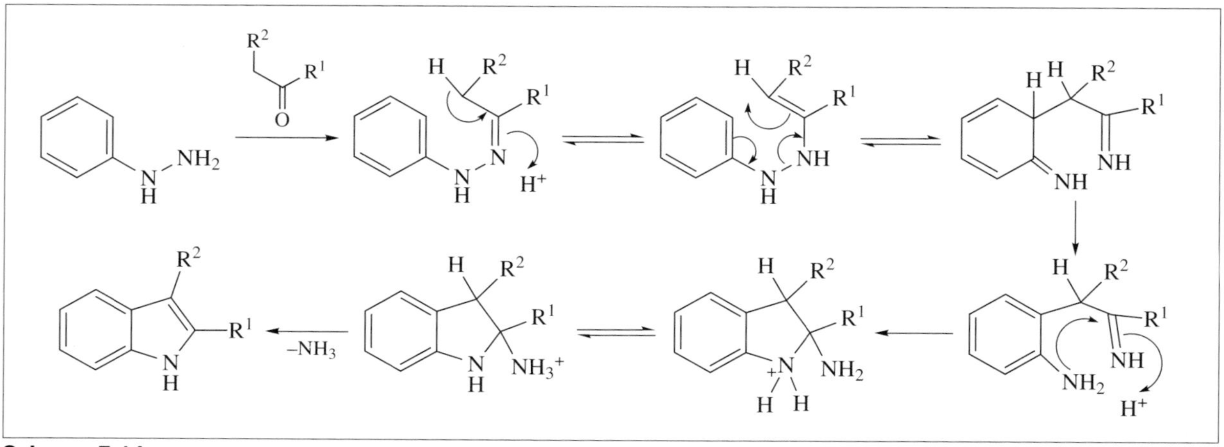

Scheme 7.14

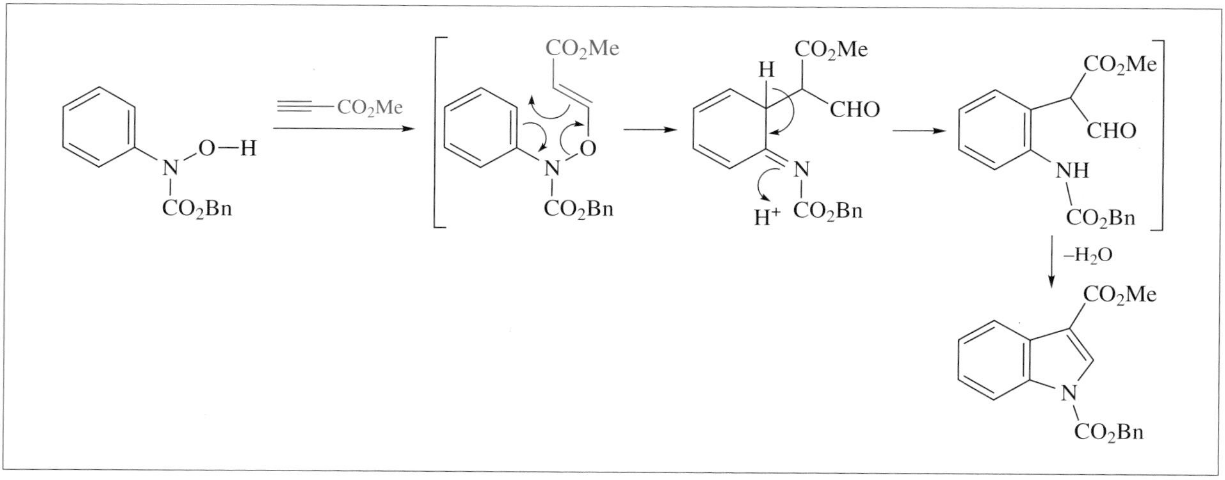

Scheme 7.15

Wender Synthesis

There are numerous other syntheses of indoles, and a modern example is the **Wender synthesis**. Here a 2-bromo-*N*-(trifluoroacetyl)aniline in THF is deprotonated by butyllithium and then, in the same pot, reacted with *tert*-butyllithium to effect halogen–metal exchange to give the dilithiated derivative. To this intermediate is added an α-bromo ketone. A carbon–carbon bond is established first between the reactants, and then cyclization occurs to form a hydroxyindoline. Finally, dehydration generates the indole (Scheme 7.16).

Leimgruber–Batcho Synthesis

When indoles unsubstituted in the heterocyclic ring are required, the **Leimgruber–Batcho synthesis** can be used. In its original form the

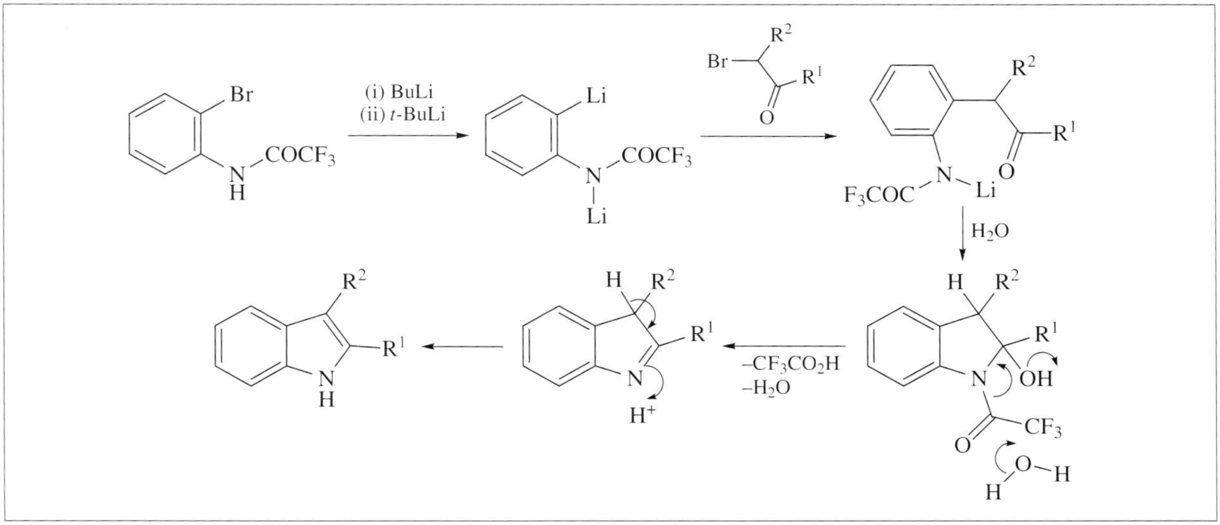

Scheme 7.16

method utilizes a 2-nitrotoluene, which when heated in a mixture containing the base pyrrolidine and *N,N*-dimethylformamide dimethyl acetal (DMFDMA) gives the corresponding β-(*N,N*-dimethylamino)-phenylethene. Reduction of this product with zinc and acetic acid forms the appropriate indole via the intermediary aniline, or its equivalent (Scheme 7.17). Cheaper reagents than DMFDMA have been recommended, including tris(piperidin-1-yl)methane, for the first step of the synthesis.

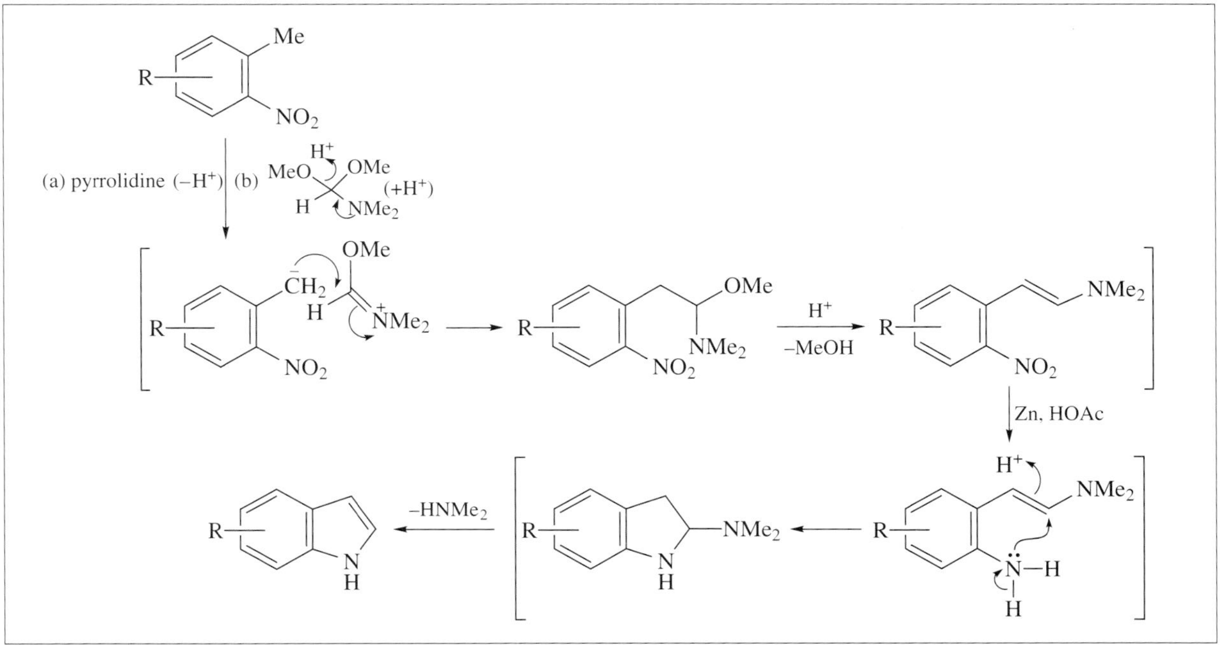

Scheme 7.17

Worked Problem 7.3

Q Suggest a synthesis of compound **7.1** from 3-bromo-4-methylaniline and cyclohexanone.

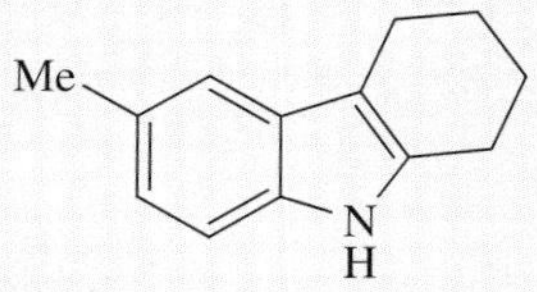

7.1

A Although it would be possible to convert 3-bromo-4-methylaniline (**7.2**) into the corresponding hydrazine, by diazotization and reduction, react it with cyclohexanone, and then subject the product hydrazone to a Fischer indolization, the bromine substituent would still remain in the indole (*note:* two isomers would form). Of course, this substituent could be displaced reductively using tributyltin hydride and a radical initiator [AIBN, azobis(isobutyronitrile)], but the overall synthesis is clumsy and non-selective and there should be a simpler route.

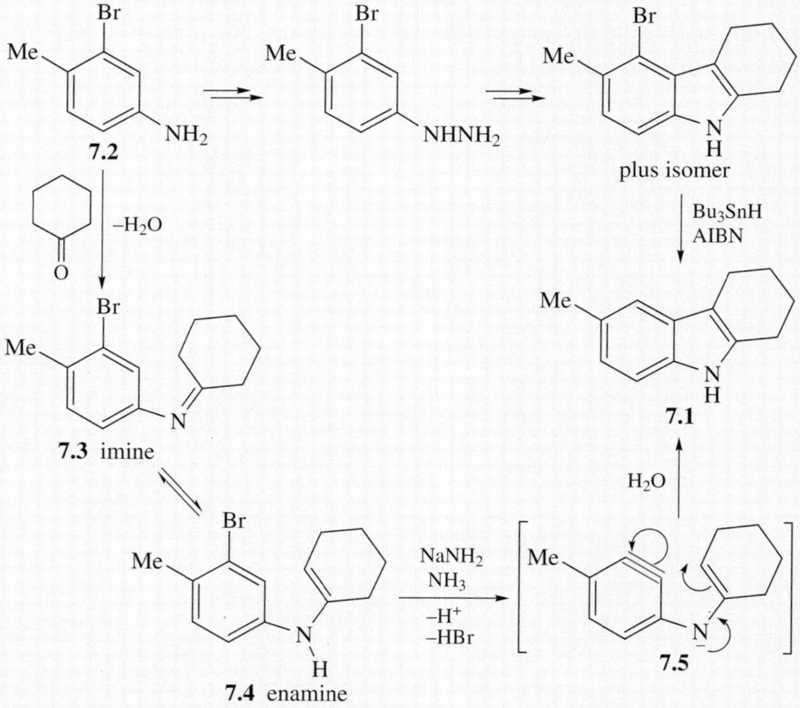

Scheme 7.18

A way forward might be to form the imine **7.3** [and hence its enamine tautomer **7.4**] by reacting the phenylamine **7.2** with cyclohexanone (Scheme 7.18). Then to generate the benzyne anion **7.5** by treating the tautomers with sodamide and sodium *tert*-butoxide in THF. Cyclization to the required indole **7.1** occurs through nucleophilic addition to the benzyne, followed by protonation during work-up.

7.1.7 Important Derivatives

Some prominent 3-substituted derivatives include skatole (3-methylindole), which has a faecal odour, and indolyl-3-acetic acid (sold as a plant rooting powder). Many indoles are biologically important; for example, tryptamine is the precursor of two hormones: serotonin, a vasoconstrictor, and melatonin, which is involved in the control of circadian rhythm. In addition, the amino acid tryptophan is an essential component of proteins (see Box 7.1).

Box 7.1 Important Indole Derivatives

Skatole

Indolyl-3-acetic acid

Tryptamine X = H
Serotonin   X = OH

Melatonin

Tryptophan

Indigo Blue

Indigo (or, more strictly, indigotin, the main component) is the blue dyestuff traditionally used to colour jeans; it and its analogues have been used by man for centuries. Indigotin occurs in plants as its precursor

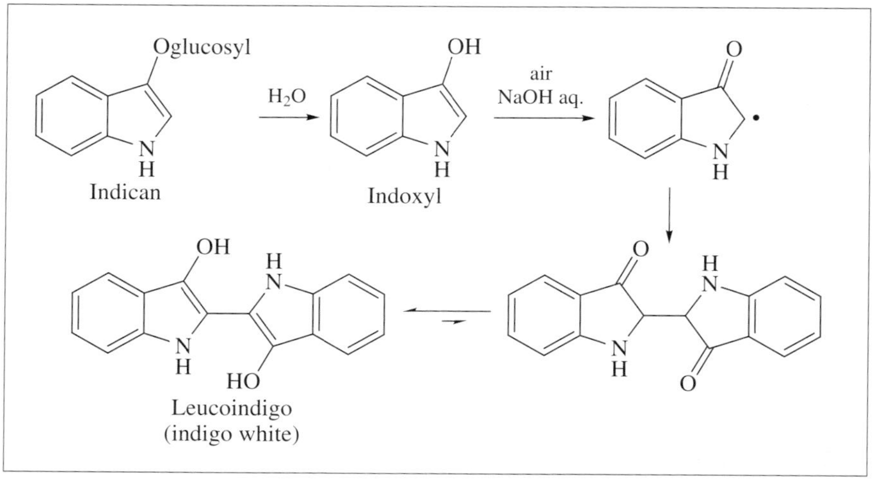

Scheme 7.19

indican (3-glucosylindole), which on hydrolysis gives indoxyl (Scheme 7.19). Indoxyl undergoes oxidative dimerization in the presence of air and aqueous alkali to leucoindigo (indigo white), and this is further oxidized to indigotin (**7.6**). Tyrian purple (**7.7**), which was used to dye the robes of Roman emperors, is 6,6′-dibromoindigo. It comes from the Mediterranean mollusc *Murex brandaris*; piles of the shells of this animal still remain outside the walls of the ancient city of Tyre.

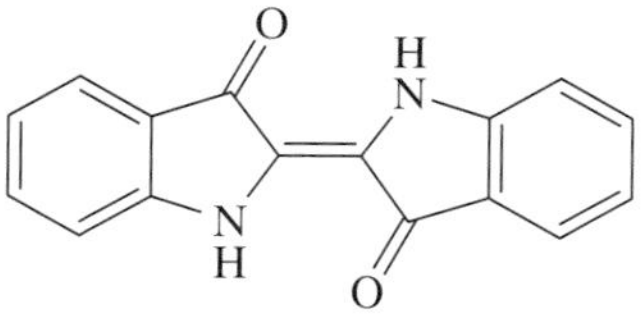

7.6 Indigotin (indigo) 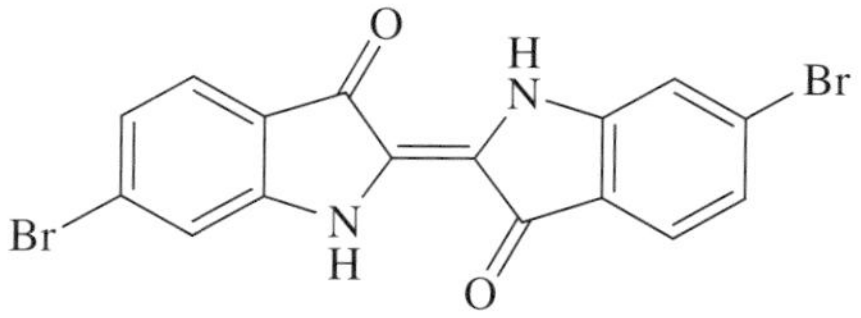   **7.7** Tyrian purple

Oxindole

Other important compounds include oxindole, which is the lactam of 2-aminophenylacetic acid. This compound has an 'active' methylene group, which can be deprotonated with a base such as sodium ethoxide, and the anion that is formed can be alkylated with a variety of electrophiles (Scheme 7.20). In the case of benzaldehydes, the initially formed aldols are unstable and these dehydrate readily to the corresponding (*E*)- and (*Z*)-(benzylidene)oxindoles.

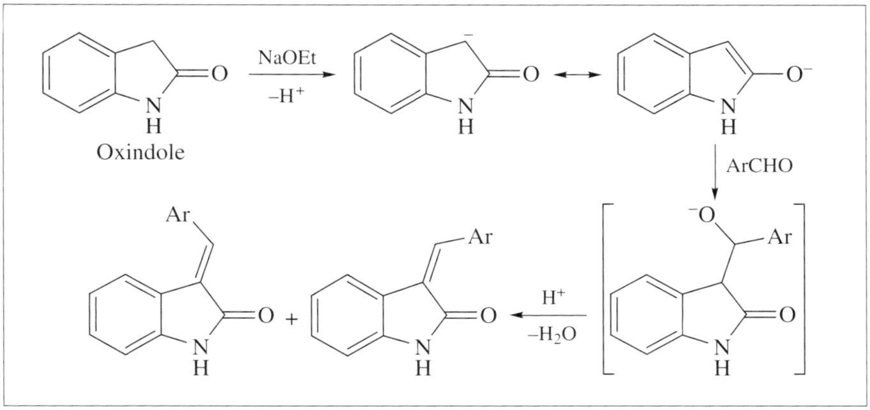

Scheme 7.20

7.2   Benzo[*b*]furan and Benzo[*b*]thiophene

7.2.1   Introduction and Reactions

These compounds are less common than indole (benzo[*b*]pyrrole). In the case of **benzo[*b*]furan** the aromaticity of the heterocycle is weaker than in indole, and this ring is easily cleaved by reduction or oxidation. Electrophilic reagents tend to react with benzo[*b*]furan at C-2 in preference to C-3 (Scheme 7.21), reflecting the reduced ability of the heteroatom to stabilize the intermediate for 3-substitution. Attack in the heterocycle is often accompanied by substitution in the benzenoid ring. Nitration with nitric acid in acetic acid gives mainly 2-nitrobenzo[*b*]furan, plus the 4-, 6- and 7-isomers. When the reagent is N_2O_4 in benzene maintained at 10 °C, both 3- and 2-nitro[*b*]furans are formed in the ratio 4:1. Under Vilsmeier reaction conditions (see Section 6.1.2), benzo[*b*]furan gives 2-formylbenzo[*b*]furan in *ca.* 40% yield.

Chlorine or bromine add across the C=C bond of the furan ring, giving 2,3-dihalo-2,3-dihydrobenzo[*b*]furans. Base-promoted dehydro-

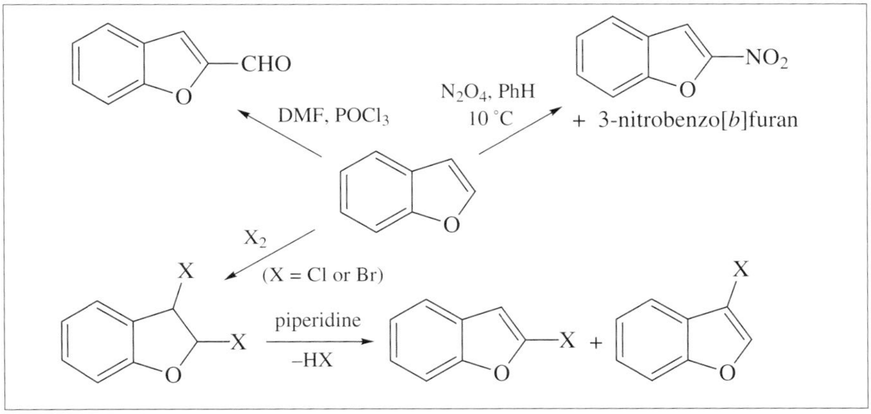

Scheme 7.21

halogenation of the dihalides affords mixtures of the corresponding 2- and 3-halobenzo[b]furans.

For **benzo[b]thiophene** the heterocycle is rather more resistant to ring opening and oxidation with hydrogen peroxide in acetic acid at 95 °C, for example, gives the 1,1-dioxide (Scheme 7.22); reduction with either sodium and ethanol or triethylsilane in trifluoroacetic acid affords 2,3-dihydrobenzo[b]thiophene. Electrophiles give mainly 3-substituted benzo[b]thiophenes, although these products are often accompanied by smaller amounts of the 2-isomers.

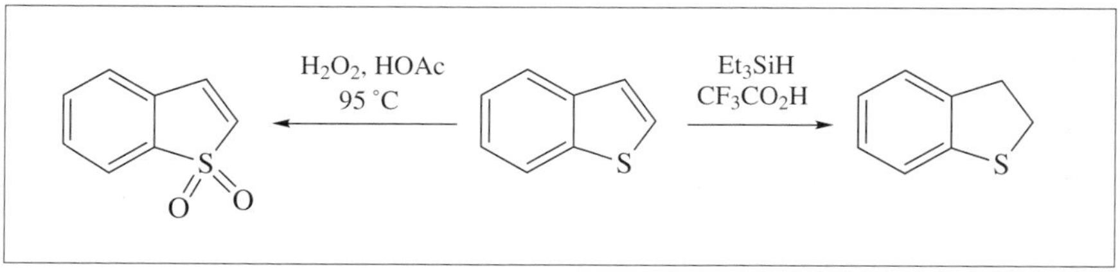

Scheme 7.22

7.2.2 Synthesis

A general route to both benzo[b]furans and benzo[b]thiophenes depends upon the cyclodehydration of either 2-hydroxy- or 2-sulfanylbenzyl ketones or aldehydes (Scheme 7.23a). 2-Acetylbenzo[b]furan can be obtained by reacting the sodium salt of 2-formylphenol with chloroacetone (chloropropanone) (Scheme 7.23b). A similar reaction using sodium 2-formylbenzenethiolate yields 2-acetylbenzo[b]thiophene.

3-Methylbenzo[b]thiophene is available through the interaction of sodium benzenethiolate and chloroacetone, followed by the cyclization of the initial product by the action of a Lewis acid, zinc chloride (Scheme 7.23c).

Scheme 7.23

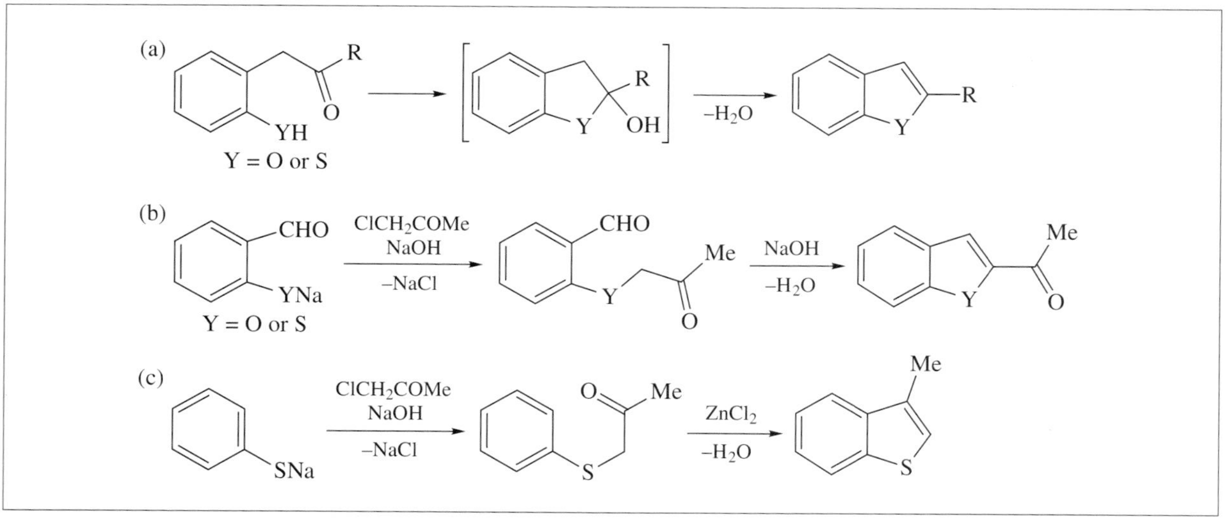

Summary of Key Points

1. Indole, unlike pyrrole, undergoes substitution at C-3 rather than C-2, even when the 3-position of the substrate is occupied. In this case the substituents at C-3 in the reaction intermediate can migrate to C-2. The ease with which they do so depends upon their ability to sustain a partial negative charge.

2. A variety of syntheses are used to prepare indoles; the most versatile for indoles substituted in the benzene ring is still the venerable Fischer indolization process.

3. Many 1- and 3-substituted indoles can be obtained from the indolyl anion; however, in order to achieve selectivity, careful attention to the counter cation is necessary.

4. 2-Substituted indoles are best obtained from 2-lithioindoles.

5. The substitution reactions of benzo[*b*]furans differ from those of indoles and *both* positions C-3 *and* C-2 can be attacked by electrophiles.

Problems

1. Why does 3-ethylindole when treated with sodium hydride and then dimethyl sulfate give a mixture of 3-ethyl-2-methylindole and 2-ethyl-3-methylindole?

2. Give practical syntheses of compounds (a)–(c).

(a)

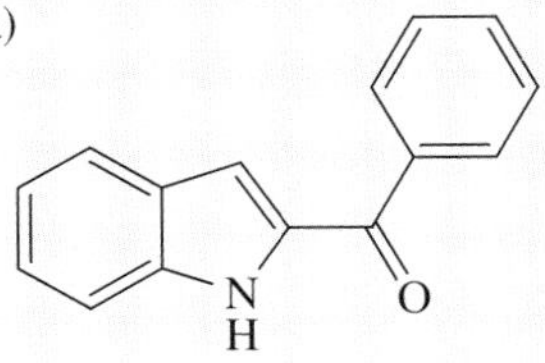

(b)

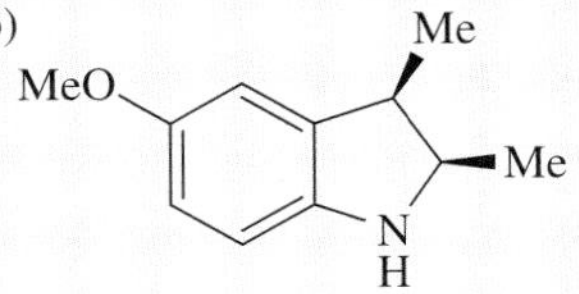

(c)

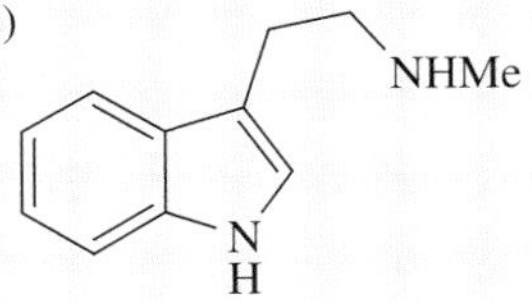

3. (a) Compare the structures of the two compounds oxindole and indoxyl; why is one more stable in air than the other? (b) If oxindole is reacted first with base, and then with 4-nitrobenzaldehyde, an adduct is obtained which dehydrates on protonation to give two isomeric products, both of which are coloured. What are the structures of these products and how are they formed? (c) Why is indoxyl unstable in air and how might you 'protect it' so as to form condensation products, similar to those formed from oxindole, with nitrobenzaldehyde and a base?

4. Devise a synthesis of 2,3-dimethylbenzo[*b*]thiophene.

Further Reading

Indoles
R. J. Sunberg, *The Chemistry of Indoles*, Academic Press, New York, 1970.
B. Robinson, *Fischer Indole Synthesis*, Wiley, Chichester, 1982.
U. Pindur and R. Adam, *Synthetic Attractive Indolization Procedures and Newer Methods for the Preparation of Selectivily Substuted Indoles*, in *J. Heterocycl. Chem.*, 1988, **25**, 1.
R. J. Sunberg, *Indoles*, Academic Press, London, 1996.
D. L. Hughes, *Progress in the Fischer Indole Reaction*, in *Org. Prep. Proc. Int.*, 1993, **25**, 607.

Benzo[*b*]furans
P. Cagniant and D. Cagniant, *Recent Advances in the Chemistry of Benzo[b]furans, Part 1, Occurrence and Synthesis*, in *Adv. Heterocycl. Chem.*, 1975, **18**, 337.

Benzo[*b*]thiophenes
B. Iddon and R. M. Scrowston, *Recent Advances in the Chemistry of Benzo[b]thiophenes*, in *Adv. Heterocycl. Chem.*, 1970, **11**, 177.
R. M. Scrowston, *Recent Advances in the Chemistry of Benzo[b]thiophenes*, in *Adv. Heterocycl. Chem.*, 1981, **29**, 171.

8

Four-membered Heterocycles containing a Single Nitrogen, Oxygen or Sulfur Atom

Aims

By the end of this chapter you should understand:

- What effect the four-membered ring has upon the chemical reactions of azetidine, oxetane and thietane, together with those exhibited by their partly unsaturated analogues
- Why β-lactams are so important
- The origins of the observed conformations of azetidine, oxetane and thietane

8.1 Azete, Azetine and Azetidine

8.1.1 Introduction

Azete, azetine and azetidine are the aza analogues of cyclobutadiene, cyclobutene and cyclobutane, respectively (Box 8.1).

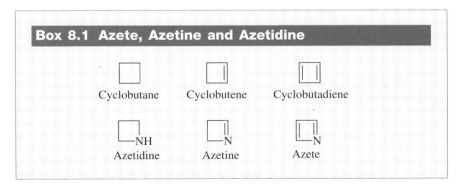

Box 8.1 Azete, Azetine and Azetidine

Cyclobutane   Cyclobutene   Cyclobutadiene

Azetidine   Azetine   Azete

The terms **azetidine** and **azetine** have been used for many years and chemists are reluctant to change them. However, the systematic name for azetidine is azetane and that for azetine is 2,3-dihydroazete.

8.1.2 Azetes

Like cyclobutadiene, azete is very unstable, but it is known in the form of certain derivatives, such as 2-phenylbenzo[*b*]azete. This compound can be synthesized in 64% yield by passing a stream of 4-phenyl-1,2,3-benzotriazoline vapour at low pressure through a tube heated to 420–450 °C. Nitrogen is lost from the starting material, possibly generating a diradical that spontaneously cyclizes to form 2-phenylbenzo[*b*]azete (Scheme 8.1).

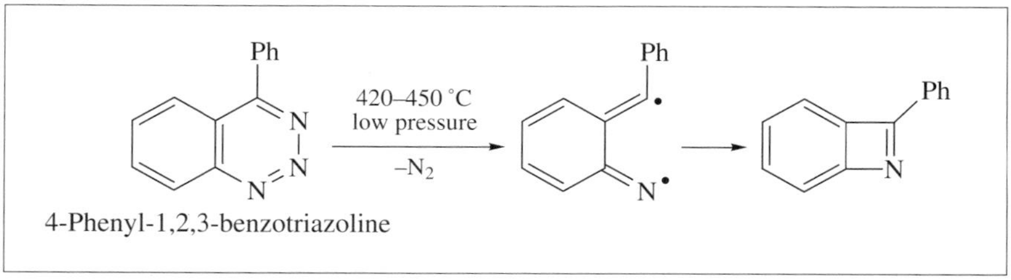

4-Phenyl-1,2,3-benzotriazoline

Scheme 8.1

As the gases emerge from the heated area of the reactor, 2-phenylbenzo[*b*]azete can be trapped out on a cold finger maintained at –78 °C. However, if the temperature is allowed to rise to about –40 °C, the compound forms a dimer.

8.1.3 Azetines

Several simple derivatives of azetine are known, although most of these bear substituents that help stabilize the compound by extending the conjugation of the endocyclic double bond. For example, 2-phenylazetine is obtained by the thermolysis of 1-azido-1-phenylcyclopropane; nitrogen is lost from the azide group and a nitrene may be formed, prior to ring expansion (Scheme 8.2).

Reduction of 2-phenylazetine with lithium aluminium hydride gives 2-phenylazetidine (Scheme 8.3); this compound and its analogues are stable compounds. Indeed, the parent heterocycle azetidine is largely unchanged when it is passed over alumina at 360 °C

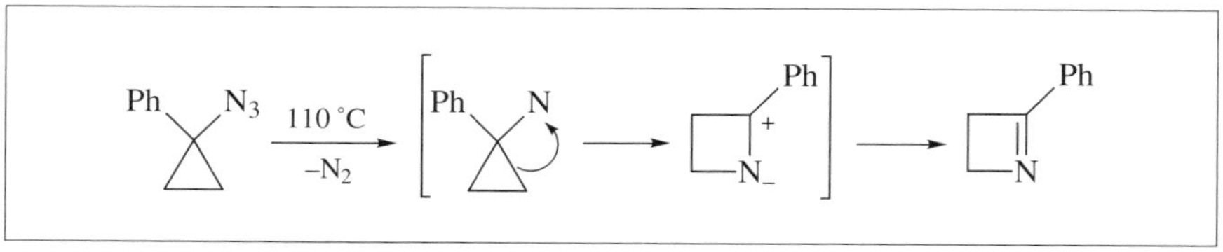

The loss of nitrogen from an azide to form a nitrene can be visualized as shown, although whether the fragmentation involves two electrons (double-headed arrows) or a single electron (fish-hooks) is not important to this discussion:

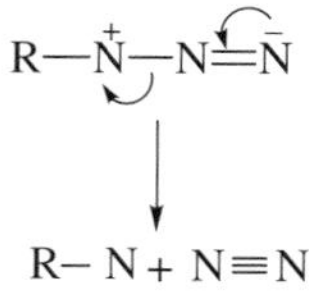

$$R—\overset{+}{N}—N\overset{\frown}{=}\bar{N}$$

$$R—N + N≡N$$

(R–N could be a singlet or triplet, depending upon R and other factors)

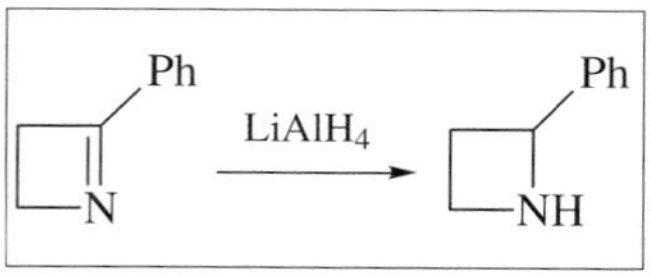

Scheme 8.2

Scheme 8.3

8.1.4 Azetidines

Azetidine, a liquid, has been known since 1899; it smells of ammonia and is a strong base. For azetidine and its derivatives the ring is puckered to reduce eclipsing interactions, the extent depending upon the substituents. The barrier to inversion is low, which helps equilibration between the two conformers (Scheme 8.4).

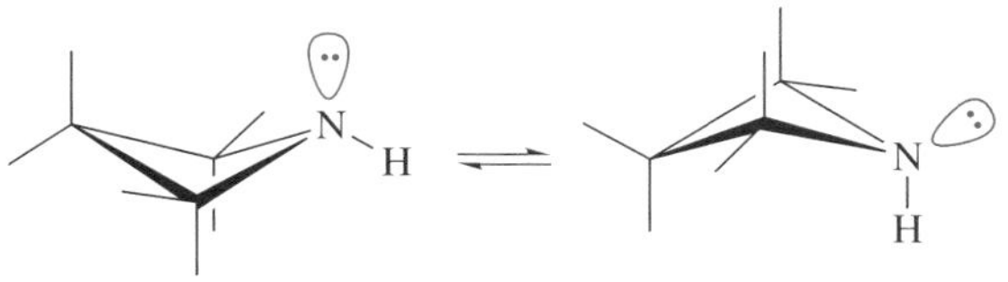

(for convenience, methylene protons are not shown)

Scheme 8.4

Like carbenes, nitrenes are electron-deficient species that contain a nitrogen atom surrounded by only six valence electrons (two are used in the R–N bond). Two types, singlets or triplets, are recognized, depending upon the electronic configuration:

Singlet nitrene

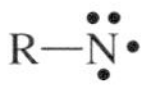

Triplet nitrene

Nitrenes react readily with electron-rich systems in order to acquire an outer electron shell on N containing eight electrons.

Synthesis

Azetidines are often synthesized by reacting 1,3-dihalogenopropanes with an amine (ammonia gives poor yields), or from propane-1,3-diamines where one N-substituent can function as a leaving group. A reaction of the latter type is used to synthesize azetidine from N,N-bis(toluenesulfonyl)-1,3-diaminopropane in two steps, the last being a reductive N-de-toluenesulfonation, caused by adding sodium to naphthalene in an inert solvent (Scheme 8.5).

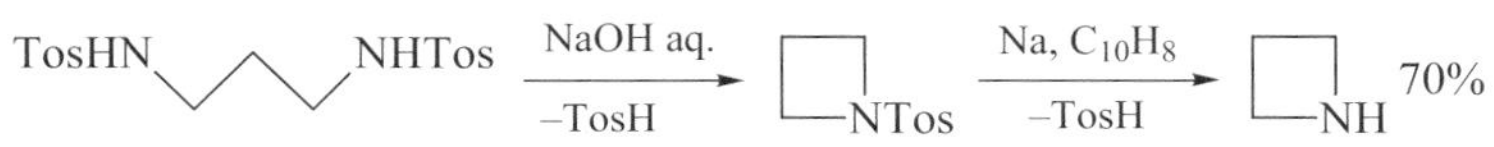

Scheme 8.5

Azetidin-2-ones (β-Lactams)

The main interest in azacyclobutanes is reserved for **azetidin-2-ones (β-lactams)**, as this ring system is found in **penicillin and cephalosporin antibiotics** (Box 8.2). These compounds are effective because the β-lactam ring is strained and readily reacts with the enzyme transpepidase, responsible for the development of the bacterial cell wall. The ring of the lactam is cleaved by this enzyme, which becomes O-acylated in the process (Scheme 8.6). Once this occurs the enzyme's normal cross-linking function is lost and the cell wall is ruptured.

Box 8.2 Penicillin and Cephalosporin Antibiotics

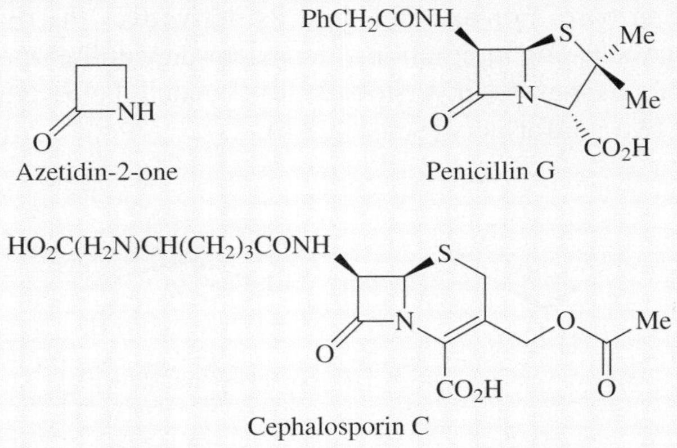

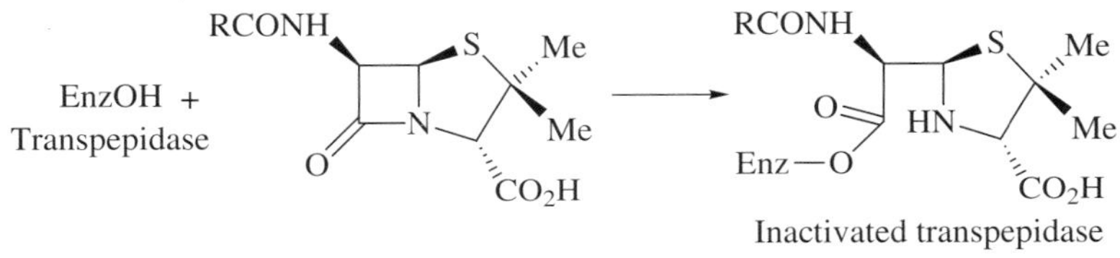

Azetidin-2-one

Penicillin G

Cephalosporin C

EnzOH +
Transpepidase

Inactivated transpepidase

Scheme 8.6

Unfortunately, many resistant bacteria have evolved which biosynthesize a lactamase enzyme (**penicillinase**) capable of cleaving the lactam ring *before* it can interact with transpepidase at the cell wall. Although cephalosporins are not inactivated by penicillinase, bacteria resistant to them are now also widespread. Indeed, the indiscriminate use of the β-lactam antibiotics in general throughout the world has severely curtailed the effective working life of these precious drugs.

Because of their immense importance to mankind, the discovery and the mode of action of the β-lactams has been surveyed in detail, and similarly there are numerous methods for their synthesis.[1]

Azetidin-2-one can be synthesized by treating 1-ethoxy-1-hydroxycyclopropane with aqueous sodium azide at pH 5.5 (Scheme 8.7a). This type of construction has wider applications and *N*-substituted derivatives are formed from 1-amino-1-hydroxycyclopropanes in two steps: first *N*-chlorination with *tert*-butyl hypochlorite [2-methylpropan-2-yl chlorate(I)], and then treatment with silver ion in acetonitrile (ethanenitrile) to release chloride ion and trigger ring expansion of the tricycle (Scheme 8.7b).

A general approach to azetidine-2-ones utilizes a cycloaddition reaction between a ketene and an imine. The ketene can be generated *in situ* by treating an acid chloride with a base, such as triethylamine [note: although these reactions are of the [2 + 2] type, *should* they be concerted they must proceed by an antarafacial, rather than a 'forbidden' suprafacial, mechanism (Scheme 8.8)].

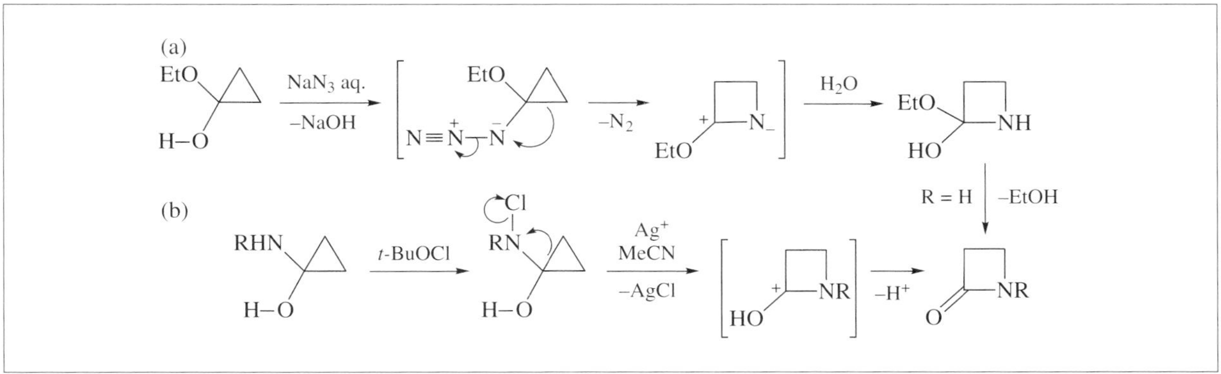

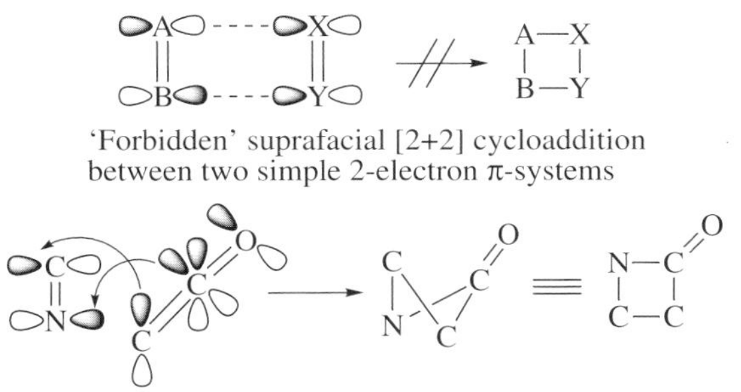

'Forbidden' suprafacial [2+2] cycloaddition between two simple 2-electron π-systems

Antarafacial alignment for an 'allowed' concerted reaction between an imine and a ketene

Scheme 8.7

A strict interpretation of the rules of orbital symmetry determines that two *simple* reactants, A=B and X=Y, each containing two π electrons, cannot combine together in a *concerted reaction* to form a four-membered saturated ring by approaching one another in the same plane (**suprafacial**). Ketenes (RCH=C=O) are linear in overall shape, but have two π-systems at a right angle to one another. This enables them to approach an imine in a 'diagonal' fashion (**antarafacial**) and react with it so that two new bonds *may* form simultaneously.

Scheme 8.8

An important point, however, is that although the configurations of the reactants are preserved in the products (*i.e.* the additions are stereoselective), some cycloadditions, including those of ketenes to imines, occur more rapidly in polar rather than in non-polar solvents (Scheme 8.9). For such examples it may well be that the addition proceeds in a *stepwise manner* (*non-concerted*), and the collapse of a dipolar intermediate is so quick that the stereochemistry of the reacting species is not compromised.

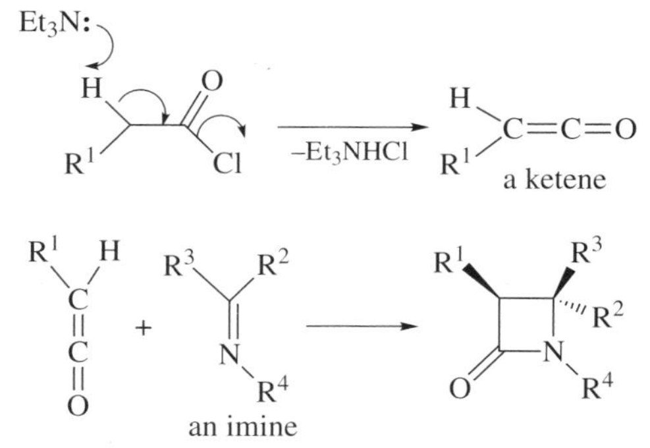

Scheme 8.9

Worked Problem 8.1

Q (i) Give a mechanism for the acid-promoted ring opening of the azetidine-2-one ring.
(ii) Suggest a reason why lactam antibiotics containing a bulky group at C-3 may act as more stable antibiotics than their unsubstituted analogues.

A The first step is protonation of the carbonyl oxygen atom, followed by the attack of water and rupture of the heterocycle (Scheme 8.10). When $R^1 = H$ the reaction with water is easier than when a large substituent blocks the approach trajectory. Consequently, 3-substituted antibiotics are more stable to ring cleavage than unsubstituted analogues. As the antibiotics are deactivated in a similar manner, 3-substituted forms are less prone to the action of a lactamase in the digestive tract.

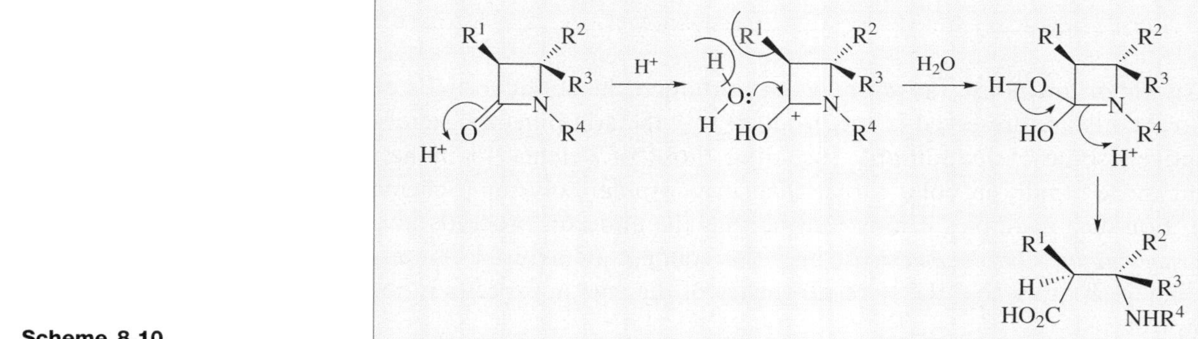

Scheme 8.10

8.2 Oxetene and Oxetane

8.2.1 Synthesis and Properties

The first reliable synthesis of an **oxetene** was that of the 2-ethoxy-4,4-di(trifluoromethyl) derivative in 1965, by reacting di(trifluoromethyl) ketone with ethoxyethyne at −78 °C (Scheme 8.11). At room temperature the product undergoes ring fission to ethyl 3,3-bis(trifluoromethyl)propenoate.

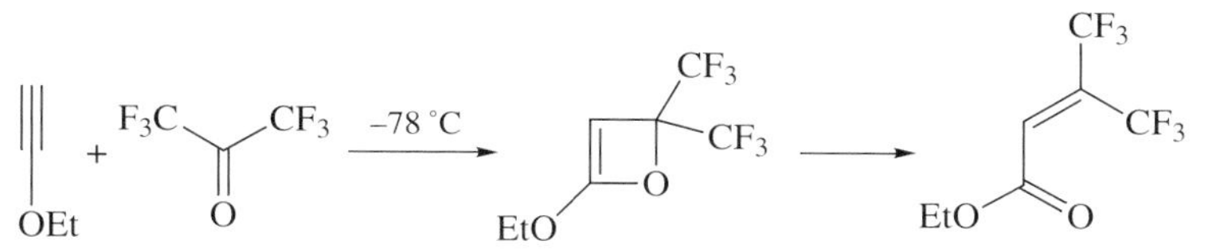

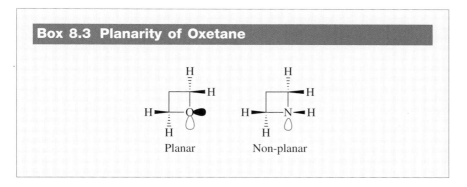

Scheme 8.11

Oxetane is a colourless liquid, formed by the cyclization of 3-acetoxy-1-chloropropane with potassium hydroxide, or from 1,3-dihydroxypropane on treatment first with sulfuric acid and then with sodium hydroxide (Scheme 8.12).

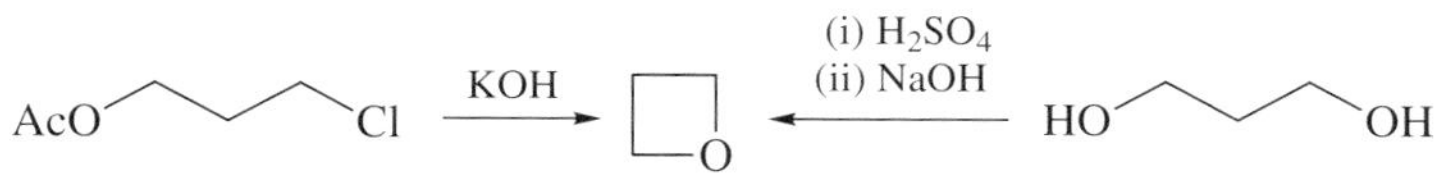

Scheme 8.12

Interestingly, the oxetane ring is essentially planar, presumably because the lone pairs on oxygen cause *fewer* eclipsing problems than a lone pair and an N–H bond do in the non-planar aza equivalent azetidine (Box 8.3).

Box 8.3 Planarity of Oxetane

Planar   Non-planar

8.2.2 Oxetanones

Oxetan-2-one (β-propanolactone) and also **oxetan-3-one** are known, although most interest lies in the former compound. It is a carcinogen, possibly because it can alkylate guanidine, one of the constituent bases of nucleic acids. On standing, oxetan-2-one slowly polymerizes, and on pyrolysis it forms propenoic acid (Scheme 8.13). Alkaline hydrolysis gives a salt of 3-hydroxypropanoic acid.

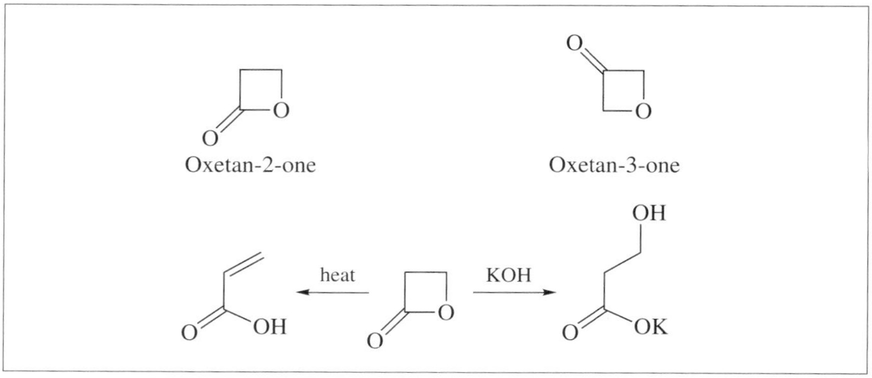

Oxetan-2-one   Oxetan-3-one

Scheme 8.13

Oxetan-2-one can be formed by treating 3-iodopropanoic acid with moist silver oxide (Scheme 8.14a), and commercially it is obtained by the cycloaddition of formaldehyde (methanal) and ketene (ethenone). At 0–20 °C, ketene undergoes self-dimerization to give 4-methyleneoxetan-2-one (Scheme 8.14b).

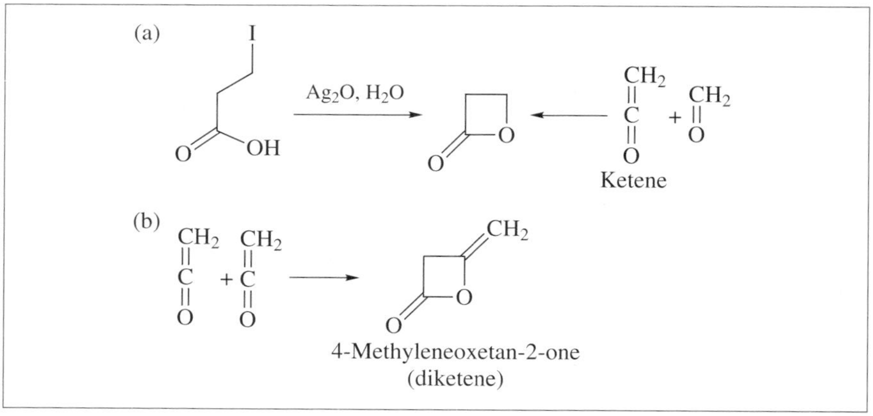

(a)

(b)

4-Methyleneoxetan-2-one
(diketene)

Scheme 8.14

8.3   Thietene and Thietane

Thietene is also called **thiete**. It can be synthesized by treating methane-sulfonyl chloride (mesyl chloride) with trimethylamine and trapping the product with *N,N*-dimethylaminoethene (Scheme 8.15). This gives 3-(*N,N*-dimethylamino)thietane 1,1-dioxide, which can be reduced by lithium aluminium hydride to 3-(*N,N*-dimethylamino)thietane, prior to *N*-alkylation with methyl iodide and base-promoted Hofmann-type elimination of trimethylamine. Reduction of thietene with sodium borohydride affords **thietane**.

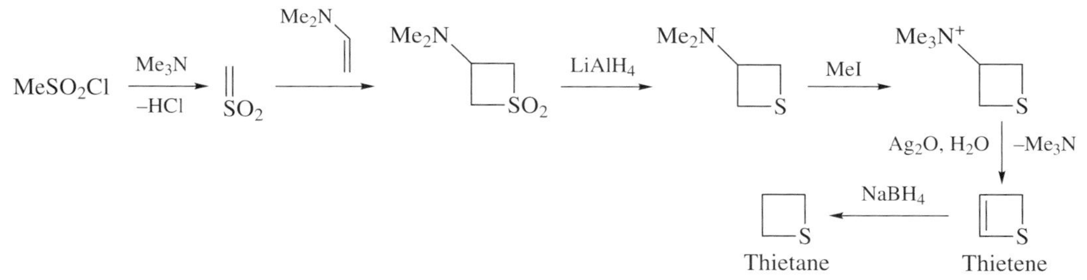

Scheme 8.15

There is theoretical interest in thietene because if it were to lose hydride ion a potentially aromatic cation having $4n + 2$ electrons, where $n = 0$, might form (Scheme 8.16). Intriguingly, the mass spectrum of thietene, molecular mass 72, shows a fragment ion peak at *m/z* 71!

Thietene is a liquid that polymerizes within an hour at room temperature. However, thietane, also a liquid, is more stable; it exists as a puckered structure, similar to that of azetidine (note: the sulfur atom and hence its lone-pair electrons occupy more space than those of oxygen).

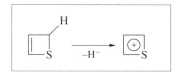

Scheme 8.16

Summary of Key Points

1. The mono aza, oxa and thia analogues of cyclobutane are all known, although their didehydro forms are unstable.

2. The spatial requirement of the hetero atom (NH in the case of azetidine) in the fully reduced forms determines the conformation of the ring.

3. The β-lactam (azetidin-2-one) unit is present in many important antibiotics.

4. Cycloaddition reactions are important in the synthesis of many of these heterocycles and they are stereoselective.

Problems

1. Suggest syntheses for compounds (a)–(c):

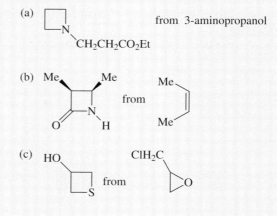

(a)   from 3-aminopropanol

(b)   from

(c)   from

Reference

1. For an introduction to this vast topic, see A. F. Pozharskii, A. T. Solatenkov and A. R. Katritzky, *Heterocycles in Life and Society*, Wiley, Chichester, 1997.

Further Reading

S. Patai (ed.), *The Chemistry of Ketenes, Allenes and Related Compounds*, Wiley-Interscience, Chichester, 1980.

Recent Advances in the Chemistry of Lactam Antibiotics, Royal Society of Chemistry, London, 1981.

E. H. Flynn (ed.), *Cephalosporins and Penicillins: Chemistry and Biology*, Academic Press, New York, 1983.

A. Hassner (ed.), *Small Ring Compounds, Part 2*, Wiley-Interscience, New York, 1983.

M. Regitz and U. Bergsträsse, *Four-Membered Rings with One or More Heteroatoms*, in *Science of Synthesis*, vol. 9, ed. G. Maas, Thieme Stuttgart, 2001, chap. 9.8.

E. Block and R. J. Lindermanin, in *Four Membered Rings with One Oxygen, Sulfur or Selenium Atom*, in *Rodd's Chemistry of Carbon Compounds*, 2nd edn., 2nd suppl., vol. IVa, chap. 1d, ed. M. Sainsbury, Elsevier, Amsterdam, 1997.

Answers to Problems

Chapter 1

1.1. (a) 2,3-Dihydro-1*H*-pyrrole; (b) 2,5-dihydro-1*H*-pyrrole; (c) 2*H*-pyran; (d) 4*H*-pyran; (e) benzo[*b*]thiophene; (f) benzo[*c*]thiophene.

1.2. (a) Antiaromatic and unstable 4*n* type; (b) aromatic, planar (4*n* + 2) type; lone pair formally on N delocalized into ring; (c) aromatic, planar and 'benzene-like constitution'; (d) Non-aromatic, non-planar aminotriene; (e) Non-aromatic, non-planar.

1.3. Normally an *N*-substituent of a piperidine larger than a proton assumes an equatorial orientation. Thus, in the isomers **A** and **B** the *transoid* conformation **A** is more stable, and hence more highly populated, than **B**.

Chapter 2

2.1. The loss of chloride ion from the adduct formed between 2-chloropyridine and a nucleophile is fast and irreversible, whereas a similar attack upon pyridine itself would require the loss of hydride ion.

2.2. Both reactions are directed to C-2(6) and C-4 (attack at C-2 is shown below). Electrophilic reactions require the loss of a proton, whereas those for nucleophiles require the loss of hydride ion. 2(4)-Halogenopyridine *N*-oxides are good substrates for nucleophilic substitution, and the thus site of attack is strongly influenced by the position of the halogen atom.

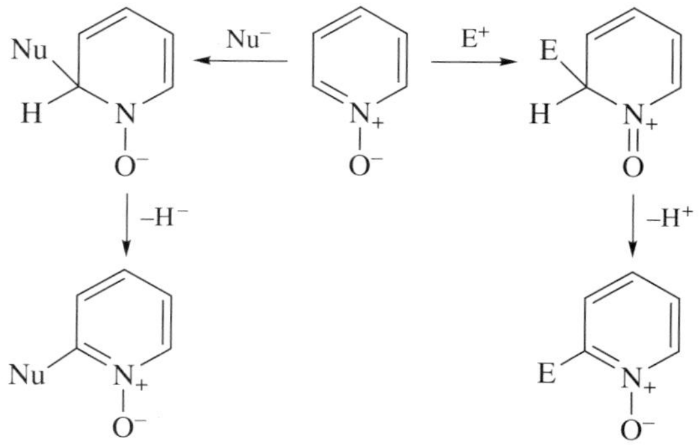

2.3. The reaction of sodamide in liquid ammonia (a strong base) causes the formation of 3-pyridyne. The subsequent addition of ammonia (or the equivalent two-step addition of amide ion and then a proton) to this compound is unspecific and leads to a mixture of 3- and 4-aminopyridines.

2.4.

(a)

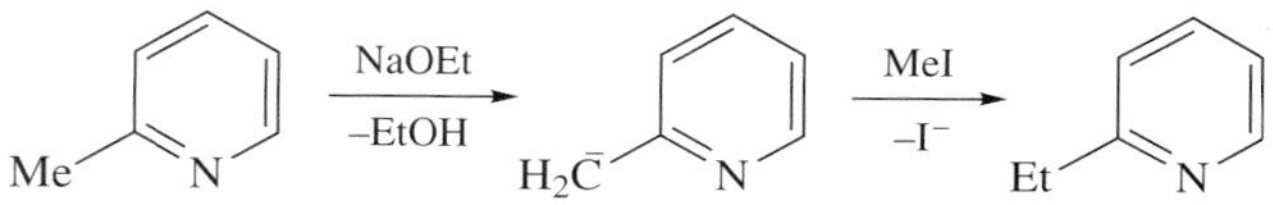

(b)

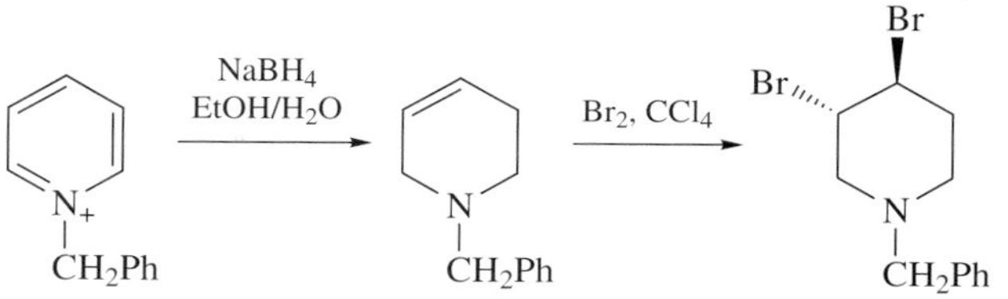

2.5. The reaction is a modified Hantzsch procedure in which ethyl acetoacetate (3-oxobutanoate) is pre-treated with benzaldehyde, *before* methyl acetoacetate and ammonia are added:

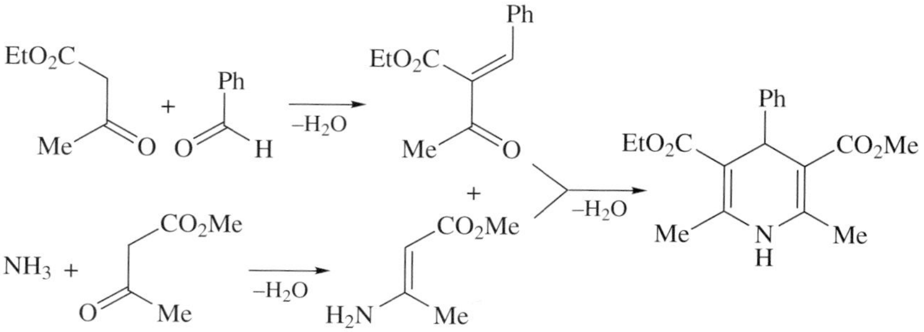

2.6. Assuming isomerism occurs to give the most stable (most conjugated and most highly substituted) dienes during the reactions, the two piperidines would undergo Hofmann exhaustive methylation and subsequent ozonolysis as shown below. The different orientation patterns are then reflected in the constitutions of the aldehydes and ketones that are formed.

(a)

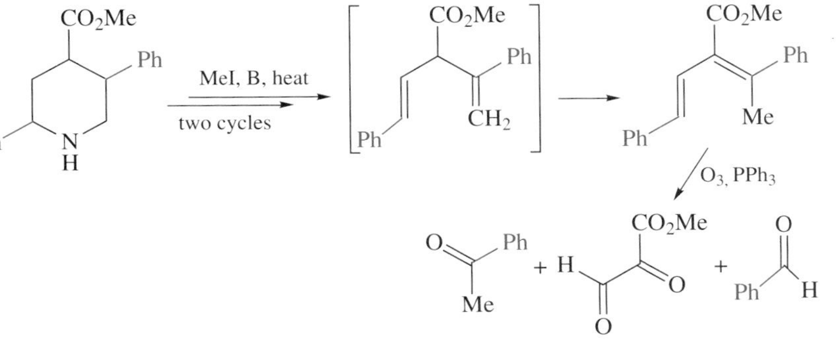

(b)

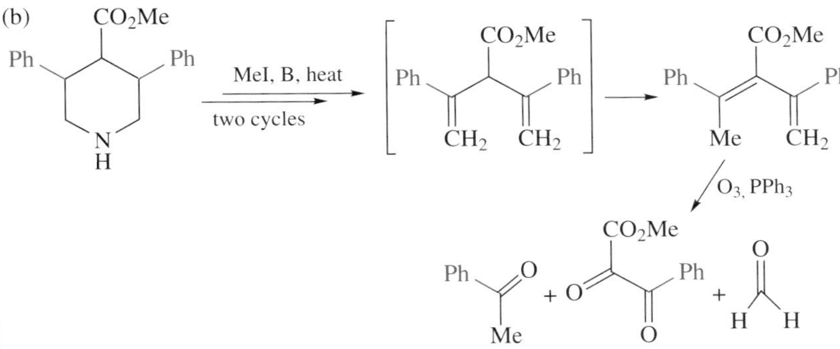

Chapter 3

3.1. The large sulfonic acid group at C-8 in quinoline-8-sulfonic acid causes an adverse steric interaction with the lone pair on the nitrogen atom. The fact that sulfonic acids of the pyridine and benzopyridine series are strong acids and self-protonate at nitrogen does nothing to solve this; however, when protonation occurs at C-8 the tetrahedral intermediate is less crowded. The formation of this intermediate and subsequent heating aids desulfonation by reversing the usual sulfonation mechanism. At this elevated temperature, sulfonation then takes place at C-6, giving the thermodynamically more stable (less crowded) quinoline-6-sulfonic acid:

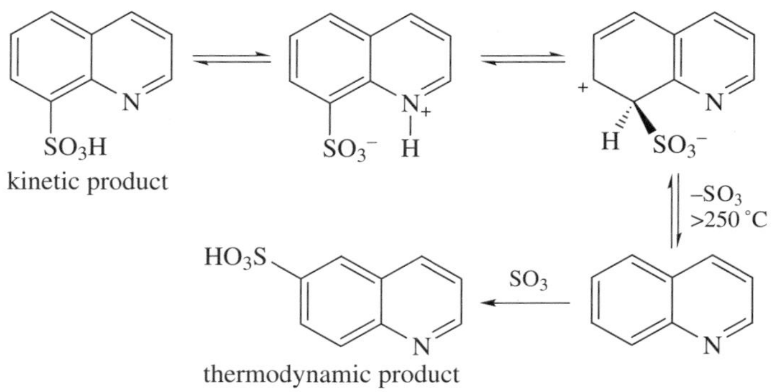

kinetic product

thermodynamic product

3.2. By reacting quinoline first with benzoyl chloride and then with aqueous potassium cyanide, the corresponding Reissert compound is obtained. The reaction is normally carried out in one operation using a two-phase system plus a phase transfer catalyst. The N-benzoyl group can then be removed by heating with sodium hydroxide solution, and the reaction is continued to hydrolyse the nitrile function to the acid:

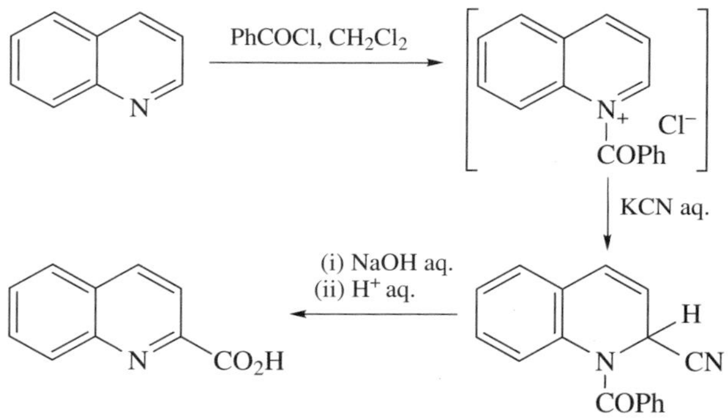

3.3. (a) Hantzsch synthesis:

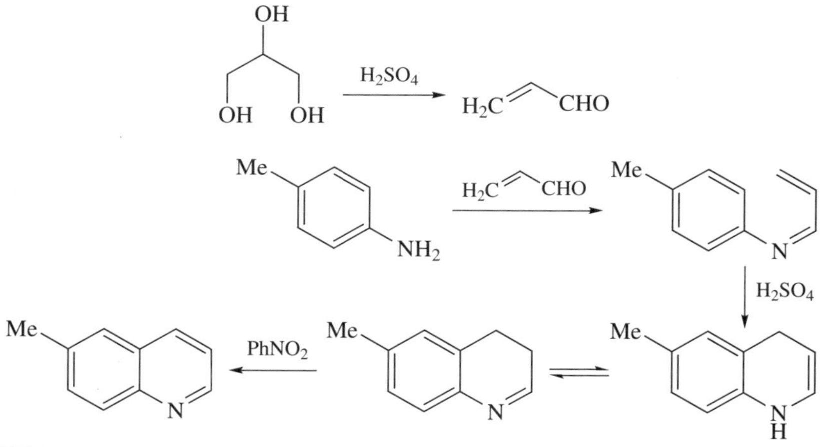

(b) Friedländer synthesis:

(c) Konrad–Limpach–Knorr synthesis (first step under thermodynamic control):

3.4. Bischler–Naperialski synthesis (used in penultimate step):

Chapter 4

4.1.

4.2.

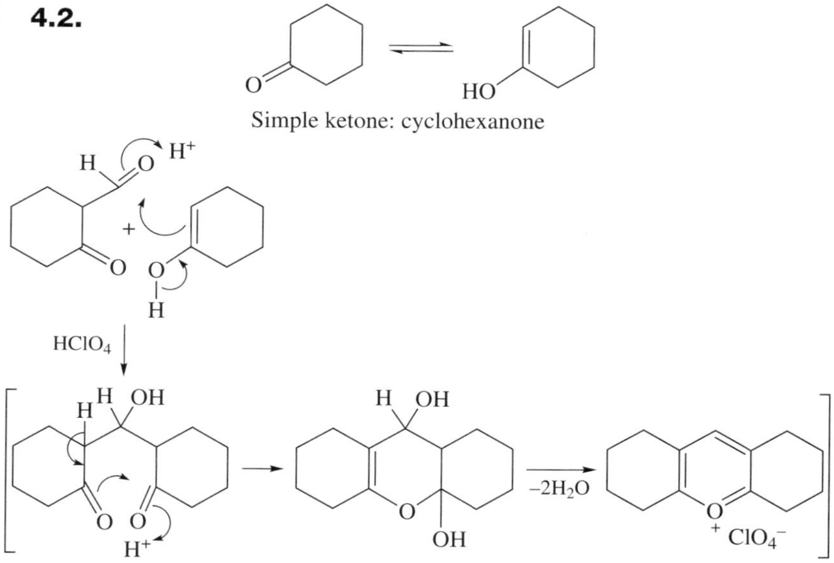

Simple ketone: cyclohexanone

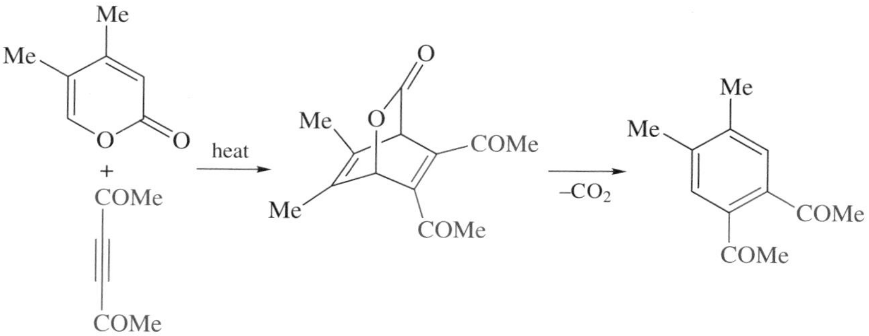

HClO$_4$

$-2H_2O$

4.3.

heat

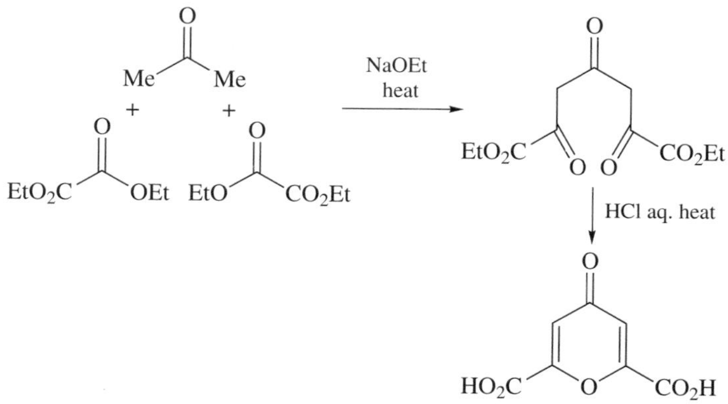

$-CO_2$

4.4.

NaOEt
heat

HCl aq. heat

Chapter 5

5.1. (a)

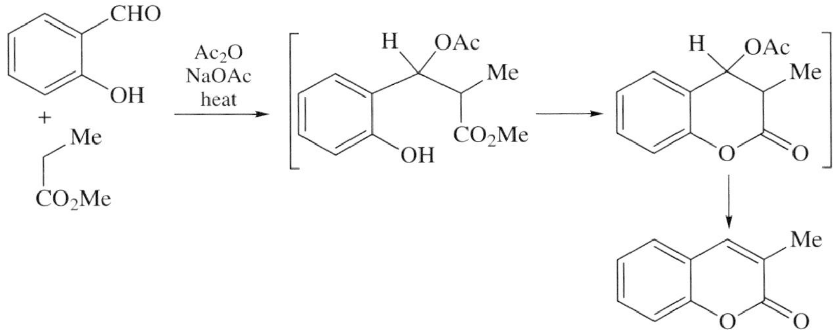

(b) In the first step of this synthesis, triethyl orthoformate (triethoxymethane) + HCl is the formylating agent; alternatively, ethyl formate (methanoate) and sodium could be used.

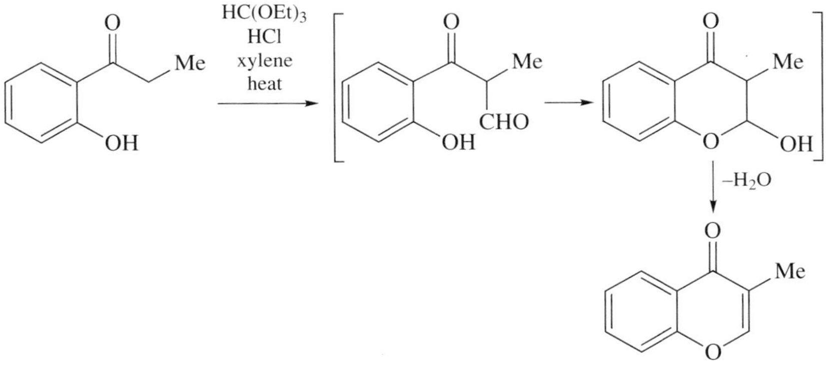

5.2.

(a)

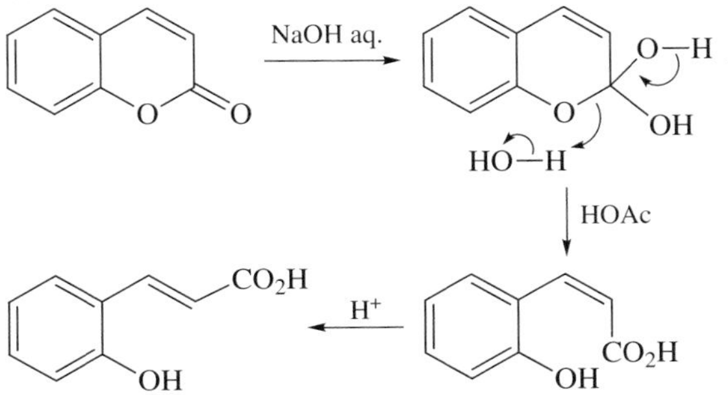

(acid-catalysed isomerism)

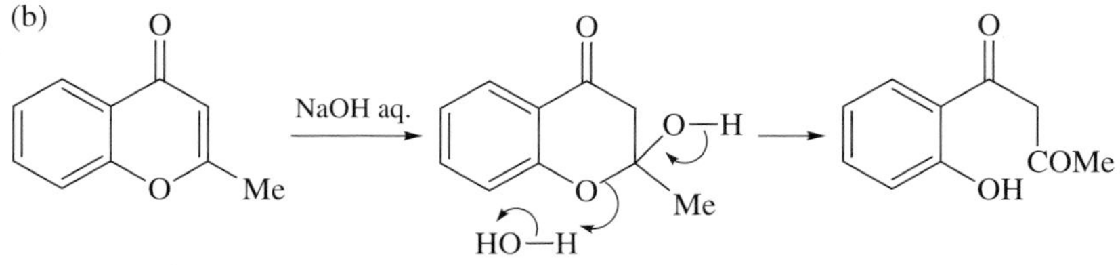

(b)

6.1. Although 2-substitution leads to a better delocalization of charge in the intermediate than reaction at C-3, here the size of the *N*-substituent dictates a reversal of the normal process (there is also the possibility that silicon may help stabilize a positive charge at a position β to it):

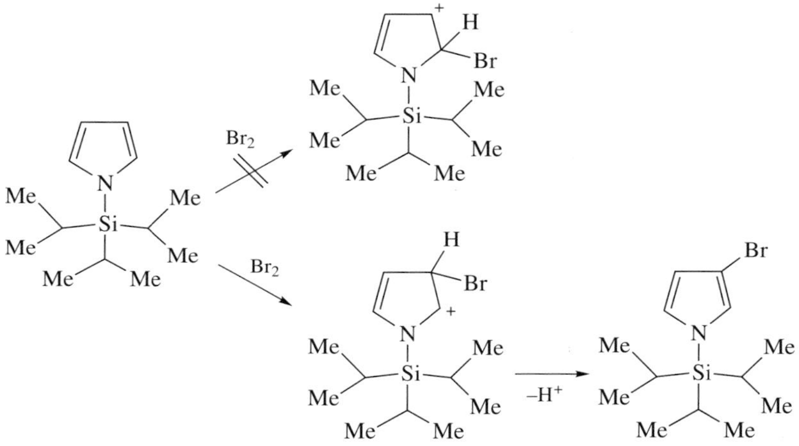

6.2. Although 2-protonation of pyrrole is favoured, there is sufficient of the 3-protonated form to initiate the addition of hydroxylamine. Once this occurs, ring opening may follow, as shown below. Subsequent addition of a second equivalent of hydroxylamine and elimination of ammonia (as ammonium chloride) then gives the dioxime.

6.3. (a) This is a classical Knorr synthesis; in fact this particular product is called Knorr's pyrrole.

(b) The starting compound is cyclohexanone, here shown as its enamine with pyrrolidine. After alkylation with chloroacetone (chloropropanone) and hydrolysis, the product diketone is cyclodehydrated by treatment with P_2O_5:

(c) 2-Bromothiophene is subjected to lithium–halogen exchange and acylated with benzoyl chloride:

6.4. The final product is furoin and it is generated by the interaction of the rearranged anion of the cyanohydrin of furfuraldehyde and furfuraldehyde itself. In aqueous medium the cyanohydrin anion would simply *O*-protonate to give the cyanohydrin, but here it does not have this option and it undergoes a protropic shift to a resonance-stabilized isomer (see below). Once this has reacted with more aldehyde, the intermediate formed also undergoes proton migration, and so permits irreversible elimination of cyanide ion, giving furoin.

Converting furfuraldehyde into its 1,3-dithian-2-yl derivative by reacting it with propane-1,3-dithiol could also be the basis of a route to furoin. Once deprotonated, this forms an 'acyl anion equivalent' that could be reacted with a second equivalent of the aldehyde and then protonated and deprotected to yield furoin:

Chapter 7

7.1. Indole reacts preferentially with electrophiles at C-3, as does its sodium salt; in this case, although an ethyl group already occupies the 3-position of the starting compound, the incoming methyl group is also bonded at this site in the initial reaction intermediate. Thereafter, as ethyl and methyl groups have similar abilities to migrate, both can move to C-2 and a mixture of products forms:

7.2. (a):

(b) Fischer indole synthesis, followed by reduction:

Note: attack on opposite face to C-3 methyl group

(c):

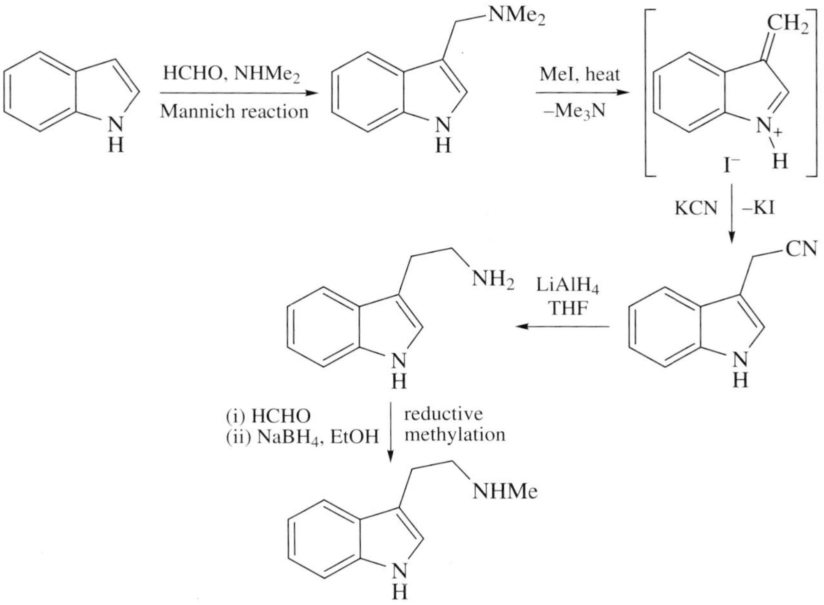

7.3. (a) Oxindole is a cyclic amide (lactam), but its ring size tends to give it rather more ketone character than is the case for acyclic amides, and on treatment with a base an anion is formed by deprotonation at C-3. This is stabilized not only by the carbonyl group (as an enolate) but also by the adjacent benzene nucleus. (b) When reacted with a carbon electrophile the benzylic position is the locus of attack, and with *p*-nitrobenzaldehyde the initial aldol-like product is easily dehydrated to a mixture of (*E*)- and (*Z*)-benzylidene derivatives (a, below). Both are highly conjugated and hence are coloured. (c) Although indoxyl also contains an 'active methylene' group and once deprotonated by base will react with benzaldehydes at C-2 in much the same way, the anion is very reactive and air must be excluded from the reaction vessel. The reason is that the indoxyl anion readily oxidizes to a resonance-stabilized radical (b, below), which couples with itself and undergoes further oxidation to give indigotin. A convenient method of 'handling' indoxyl is to use it in the form of its 1,3-diacetyl compound (c, below), a stable crystalline solid.

(a)

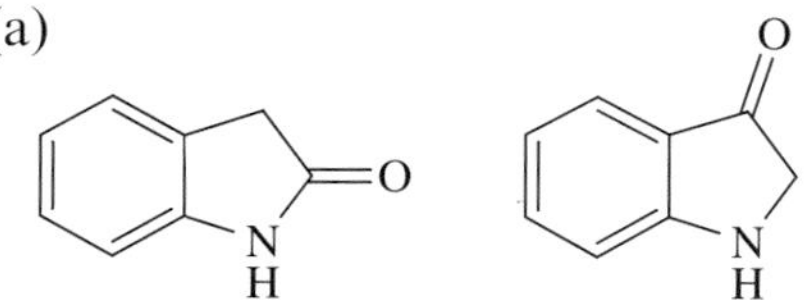

Oxindole   Indoxyl

(b)

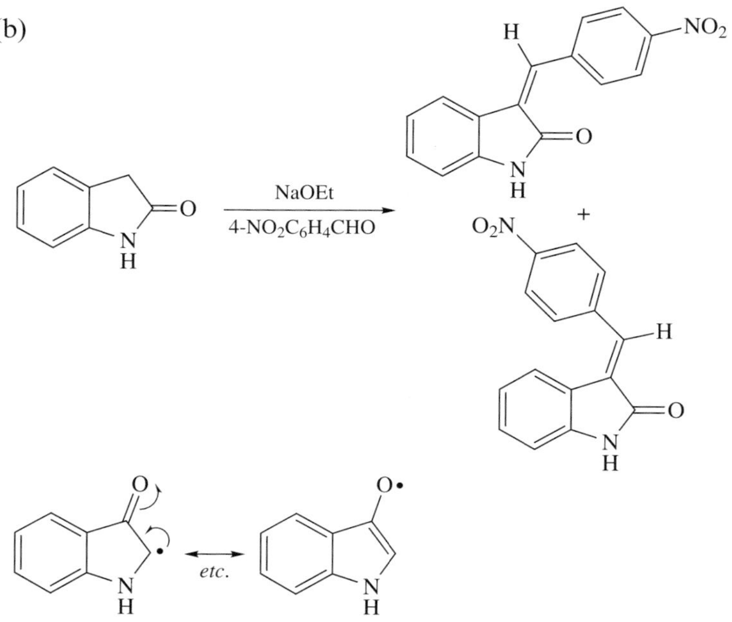

NaOEt
4-NO$_2$C$_6$H$_4$CHO

etc.

(c)

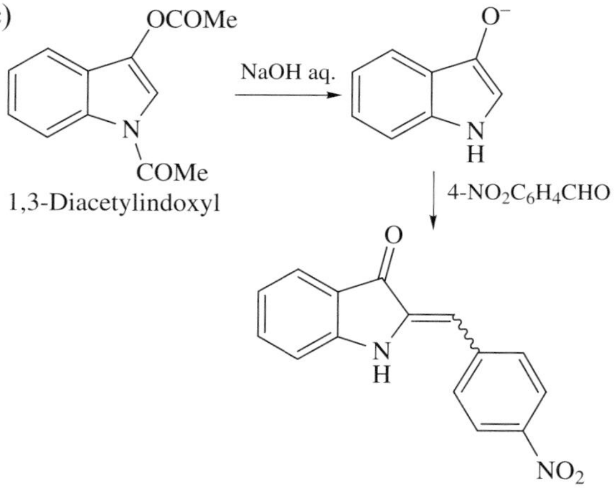

1,3-Diacetylindoxyl

NaOH aq.

4-NO$_2$C$_6$H$_4$CHO

7.4. Several disconnections are possible for this compound, of which (a) and (b) are fairly obvious. However, although the starting material for (a) could be obtained from sodium benzenethiolate and 3-chlorobutan-2-one, that for (b) is more difficult as *ortho* alkylation of benzenethiol would be necessary:

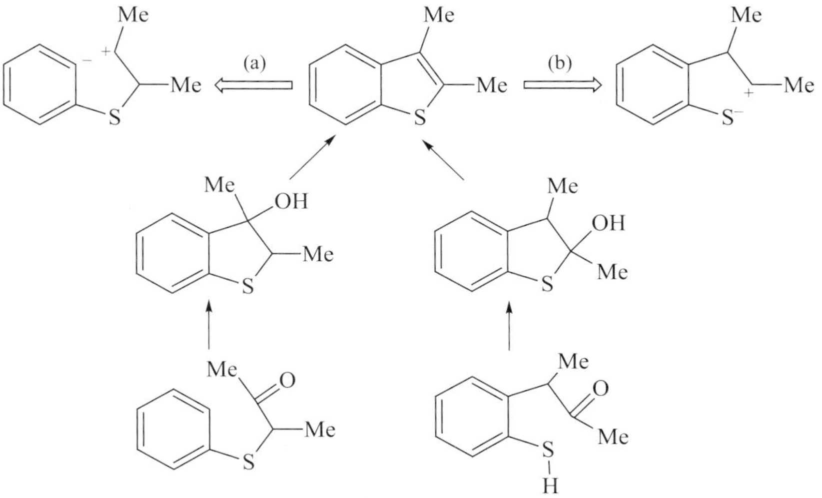

An alternative would be to establish the aryl side chain selectively by a rearrangement process:

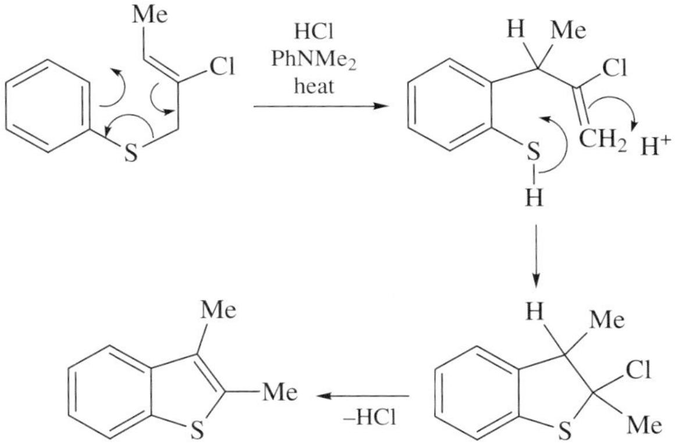

Chapter 8

8.1. (a) The synthesis begins with a conjugative addition of 3-aminopropanol to ethyl acrylate, forming an adduct that can be cyclized by treatment with thionyl chloride and then with a base, sodium carbonate:

(b) The *cis* disposition of the two methyl groups suggests a cyclo-addition approach involving (*Z*)-but-2-ene as one reaction component. The other is chlorosulfonyl isocyanate (used because it is very reactive and because the chlorosulfonyl group can be easily removed at the end of the reaction). Whether the first stereoselective reaction is an antarafacial [2 + 2] concerted reaction or a stepwise process with a very fast collapse of the dipolar intermediate is arguable. Certainly the rate of the addition is greater in polar solvents than in non-polar ones.

(c) If the epoxide is treated with hydrogen sulfide, 3-chloro-2-hydroxypropanethiol is obtained. This compound is cyclized to 3-hydroxythietane by exposure to aqueous base (barium hydroxide solution was used):

Subject Index

Aminopyridines 34
Aminoquinolines 48
Anomeric effect 12
Anthocyanidins 68
Anthocyanins 68
Anti-aromaticity 8
Aromaticity 6
Azepines 4
Azete (azacyclobuta-
    diene) 9, 115
Azetidine 115
Azetidin-2-ones 117
Azetine 115

Benzo[*b*]furan 111
Benzopyridines 42
Benzopyrylium salts 68
Benzo[*b*]thiophene 111
Bischler–Napieralski
    reaction 52, 129

Carbazoles, tetrahydro-
    98
Carbohydrates 65
Cephalosporins 117
Chelidonic acid 67
Chichibabin reaction 24
Chlorophyll a 5
Chromones 68, 72
Cocaine 6
Conformation 10
Conrad–Limpach–Knorr
    synthesis 49
Coumalic acid 63
Coumarins 68, 70
Cyanidin 69

Cyanin 68
Cyclooctatetraene 9

Dihydropyrans 58

Fischer indole synthesis
    105, 138
Flavones 68
Flavylium salts 68
Friedländer synthesis
    47, 129
Furan 85
    cycloaddition reactions
    88
    electrophilic substitu-
    tion 86
    metallation 87
    synthesis 89
Furfural (2-formylfuran)
    90

Guareschi synthesis 28

Haem 4
Halopyridines 20, 25
Hantzsch synthesis 28,
    126, 128
Hantzsch–Widman sys-
    tem 3
Hard/soft notation 62
Heroin 6
Hofmann exhaustive
    methylation 38, 127
Hückel's rule 7, 97
Hund's rule 8
Hydroxypyridines 32

Indican 109
Indigotin (indigo blue)
    5, 109
Indigo white 110
Indole 97
    anion 103
    reactions 98, 135
    reduction 104
    ring expansion 104
Indole-3-acetic acid 109
Isoflavones 69
    reactions 51
    reduction 51
    structure 51
    synthesis 52
Isoquinoline 4

Knorr synthesis 83, 133
Kojic acid 63
Kolbe reaction 81

β-Lactams 117
Leimgruber–Batcho
    reaction 106

Mannich reaction 74,
    101
Melatonin 109
Methylpyridines 29
Methylquinolines 47
Morphine 6

Nicotinamide adenine
    dinucleotide (NAD^+)
    36
Nicotine 6

Nicotinic acid 30
Nifedipine 6
Nomenclature 1
Non-aromaticity 8

Oxetane 121
Oxetanones 122
Oxetene 121
Oxindole 110, 137

Paal–Knorr synthesis 82
Paraquat 6
Penicillins 117
Picolines 29
Picolinic acids 30
Pictet–Spengler synthesis
    53
Piperidine 11, 36, 60
Pomeranz–Fritsch
    synthesis 54
Pyranones 64
Pyrans 58
Pyrans, dihydro- 58
Pyrans, tetrahydro- 3,
    13, 58, 65
Pyridine 18
    acylation 20
    addition elimination
        20
    electrophilic substitu-
        tion 19
    lithiation 28
    nucleophilic substitu-
        tion 23

reduction 36
structure 7
Pyridine N-oxides 22
Pyridines, amino- 34
Pyridines, hydroxy- 32
Pyridines, methyl- 29
Pyridinium salts 34
Pyridones 32, 51, 58, 63
Pyridoxine 6
Pyridynes 27, 126
Pyrrole 77
    anion 81
    cycloaddition reactions
        82
    reduction 84
    structure 8
    synthesis 82
Pyrrolidine 81, 84
Pyrroline 84
Pyrylium salts 58

Quinoline 42
    electrophilic substitu-
        tion 44
    structure 43
    synthesis 54
Quinolines, amino- 48
Quinolines, methyl- 47
Quinolizinium cation 42
Quinolones 48

Reissert compounds 45,
    128
Ring current 10

Ring strain 11
Rotenone 68

Saccharides 65
Seratoin 109
Skatole 109
Stilbazole 31
Strychnine 5

Tetrahydrocarbazoles 98
Tetrahydropyrans 3, 13,
    58, 65
Thietane 122
Thiophene 91
    electrophilic substitu-
        tion 91
    metallation 93
    oxidation 92
    synthesis 93
Tryptamine 109
Tryptophan 109
Tyrian purple 110

Vilsmeier reaction 74, 99

Wafarin 68
Wender indole synthesis
    106